STAR CLUSTERS

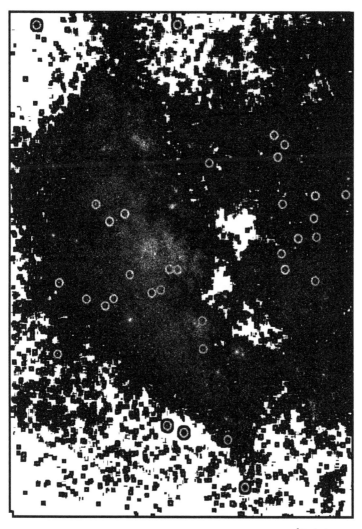

Thirty-four Globular Clusters in the Southern Milky Way.

HARVARD OBSERVATORY MONOGRAPHS

No. 2

STAR CLUSTERS

BY

HARLOW SHAPLEY

Published for the Observatory
by the

McGRAW-HILL BOOK COMPANY, Inc.
NEW YORK: 370 SEVENTH AVENUE
LONDON: 6 & 8 BOUVERIE ST, E. C. 4
1930

THE MAPLE PRESS COMPANY, YORK, PA.

PREFACE

FOR more than five years the writing of this second monograph of the Harvard Observatory series has been in progress. Several factors have contributed to the postponement of its publication, but the chief cause of delay has been the desire to provide a revised system of parallaxes for globular clusters. All considerations of the dimensions, star densities, and luminosities of clusters, and most of the conclusions concerning the dimensions of the Galaxy and the distance to external systems, depend at the present time on the period-luminosity relation of Cepheid variable stars. In the preparation of a monographic treatment of star clusters much preliminary work on Cepheids has therefore been necessary. . At the McCormick and Mount Wilson observatories investigations of proper motions have been undertaken in order to fix the zero point of the period-luminosity curve. At Harvard we have determined periods and magnitudes of many variable stars in the Large and Small Clouds of Magellan to provide a firmer photographic connection between period and luminosity.

The revision of the parallaxes of globular clusters has also depended largely on the study of individual systems. In my determination in 1917 of the parallaxes of sixty-eight globular clusters, there were only five systems for which the variables had been sufficiently studied to enter directly into the measurement of cluster distances. We now have nineteen clusters in which variable stars have been adequately studied; the periods and median magnitudes of more than five hundred individual variables have been derived. In 1917 the magnitudes of the high luminosity stars had been measured for twenty-eight globular clusters; the number is now increased to forty-eight.

Much of the recent investigation of magnitudes in clusters is the work of Miss Helen Sawyer at the Harvard Observatory; my earlier photometric studies preliminary to the revision of cluster distances were carried on at Mount Wilson and Harvard with the aid of several assistants.

Probably one of the most satisfactory features of the present volume is the bibliography in Appendix C. It is essentially complete for all papers bearing directly on star clusters published during the past fifty or sixty years. Particular attention should be directed to the treatises by ten Bruggencate and Parvulesco, who handle some special subjects in the field of clusters more fully than they are treated here.

I am indebted to Dr. Cecilia Payne for extensive assistance in the preparation of manuscript and bibliography and in other details. In the interest of this study of clusters Dr. Walter S. Adams has generously transferred to the Harvard Observatory the photographs I made with the 60-inch and 100-inch reflectors at Mount Wilson. Miss Jenka Mohr has assisted in the preparation of the manuscript and in editorial matters. To several members of the Observatory staff I am indebted for incidental or particular assistance with illustrations, computations, and routine work on manuscript and proof. My greatest debt is to the star clusters themselves, which have provided persistent excitement and inspiration.

H. S.

CAMBRIDGE, MASS,
June, 1930.

CONTENTS

STAR CLUSTERS

CHAPTER I

INTRODUCTORY SURVEY

FROM a place deeply involved in stellar organizations we look out at the sidereal universe. Immediately around our minute association of sun and planets are the bright white stars of the Ursa Major cluster, with one of its members, Sirius, but a few light years distant. Much greater than such near-by groups is the local system, a star cloud that includes thousands of the surrounding bright stars and a million or so of the fainter. Still more inclusive is the Galaxy, which holds sun, clusters, and star clouds, and which, in turn, may be but one unit in a higher example of the gravitational clustering of sidereal bodies.

1. The Significance of Clusters.—In time as well as in space we are involved in clustering tendencies—we witness the sidereal universe undergoing formative and disruptive processes. The evidence of our situation in the mid-course of the transformation of stellar systems, rather than at the beginning or end, is well illustrated in the mixture of stars, nebulae, and sidereal groups that go to make up the Clouds of Magellan or any large section of the Galaxy. In such places we find star clusters in nearly all degrees of richness and condensation; we find nebulosity associated with stars and groups of stars, and everywhere there are large irregularities in the general stellar distribution. The contrast is striking between the smoothness of a typical globular cluster, where an obvious state of equilibrium prevails, and the heterogeneity of content and irregularity of form in the galactic system. The globular cluster, when undisturbed from without, apparently attains a steady stage, but the star clouds and the Galaxy, with their clusters, star streams, and secondary organizations, are far from that condition.

By star clusters we ordinarily mean both the typical globular systems and the more numerous and less well-defined open clusters which range, for instance, from the Hyades to the fairly compact system of Messier 11. Such clusters are composed of stars which are known to be physically connected or may be assumed from their apparent positions to constitute distinct physical organizations. But also, in all parts of the sky, among the faint stars there are thousands of less obvious groupings. The studies of star distribution on the astrographic charts by Turner,[1] and by Öpik and Lukk,[2] indicate that the distribution is not at random. Working on this problem with Harvard plates,[3] I have shown that the clustering and vacancies are real and are not to be attributed to occultation by nebulosity. Random distribution among galactic stars, it seems, does not exist. The observed irregularities in the star counts, beyond those allowed by the law of chance, are to be attributed, in general, to the very prevalent stellar associations, which are not commonly recognized by casual inspection and cannot be separated from surrounding stars except through laborious investigation.

The typical star clusters, however, are in themselves numerous and widely distributed, and their problems are intimately interwoven with some of the most significant questions of stellar organization and galactic evolution. The general study of clusters deals with a wide variety of subjects. It involves, for instance, the problems of supergiant stars, stellar luminosity curves, irregularities in stellar distribution, star streaming, island universes, and the genesis of galactic systems; it considers primarily, however, the composition, structure, distribu-

---

[1] Obs., 48, 173, 1925.
[2] Publ. de l'Obs. Astr. de l'Univ. de Tartu (Dorpat), 26, No. 2, 1924.
[3] H. C 281, 1925.

tion, and cosmic position of the easily recognizable galactic and globular clusters, and in the following chapters these groups will receive almost exclusive attention.

2. Historical Notes on Clusters.—The history of the scientific study of star clusters is neither extensive nor very significant. Several clusters of naked-eye stars—for example, the Pleiades, Praesepe, Coma Berenices—have, of course, always been known, though their definite assignment to the cluster category came with the work on proper motions in the last 50 years. For a few constellations, the majority of the bright stars are now known to lie near together in space and to form physical systems. Such constellation groups are Taurus, Orion, Ursa Major, Perseus, Scorpio, Sagittarius, and Vela. But no close physical connection exists for the bright stars of Cassiopeia, Lyra, Aquila, Canis Major, and many others.

A score of the brighter galactic clusters and half a dozen of the globular clusters can be seen with the naked eye under good conditions. These were probably all known, therefore, to the ancients, but only a few appeared in our permanent records before the latter half of the eighteenth century.

The records of Hipparchus contain references to the double cluster in Perseus and to Praesepe, although neither was recognized as a group of distinct stars until the invention of the telescope. Both were first resolved by Galileo, who described "the nebula called Praesepe" as "not one star, only, but a mass of more than forty small stars."[4]

Messier 22, the first globular cluster to be recorded as such, was discovered by Ihle[5] in 1665; ω Centauri was noted as a lucid spot in the sky by Halley in 1677 and had previously been known to Bayer as a hazy star and to Ptolemy as a star in the cloud on the Horse's back; in 1702 Kirch discovered Messier 5, and the famous Messier 13 (the Hercules cluster),

---

[4] Galileo, Nuncius Sidereus, 1610; Allen, Star Names and Their Meanings, 113, 1899.
[5] Wolf, R , Geschichte der Astronomie, 420, 1877.

the brightest in the northern sky, was accidentally found by Halley in 1714.

The open cluster Messier 11 had already been recorded by Kirch in 1681; but the majority of bright galactic clusters, except the Pleiades and the Hyades, were first recorded as such by Messier[6] in 1771. The conspicuous groups of stars around η Carinae and the cluster near κ Crucis were discovered by Sir John Herschel.

For both open and globular clusters, as well as for bright nebulae of all kinds, the systematic listing by Messier in 1784 marked an epoch in the recording of observations. The Herschels advanced the work materially. Especially significant were the General Catalogue published by Sir John Herschel in 1864[7] and its important sequels by Dreyer in the New General Catalogue and the Index Catalogues.

Schultz and Barnard were among the pioneers in determining visually the positions of the individual stars in globular clusters. The superior photographic method of charting positions was first used by the Henrys and Gould for galactic clusters and by Scheiner, Ludendorff, and von Zeipel for globular systems.

The Pleiades, the Hyades, Praesepe, h and χ Persei, and some of the other bright galactic groups have, for the past 50 years or more, been the subject of frequent investigations of positions and proper motions. It is not unfair to say, however, that, except for studies of these nearby objects, the work done on individual clusters before the present century is now of little value. The development of photographic methods, the modern large telescopes with their rapid spectroscopes, and the standardizing of magnitude sequences have all tended to make the earlier work obsolete. The present views of the nature, dimensions, and significance of the globular clusters are less than 15 years old.

In striking contrast to the present conception of a hundred globular clusters and hundreds of thousands of extra-galactic

[6] Hist. de l'Acad. R. des Sci., Paris, 435, 1771.
[7] Phil. Trans , 154, 1, 1864.

nebulae spread throughout measured millions of light years, with diameters of hundreds of light years for the clusters and thousands of light years for the star clouds and nebulae, is the picture suggested by Halley's comment on his discovery of the Hercules cluster:

> But a little patch—and similar to the lucid spot around Theta Orionis [Orion Nebula] Andromeda [Andromeda Nebula], and in the Centaur [ω Centauri]—most of them but a few minutes in diameter; yet since they are among the fixed stars . . . they cannot fail to occupy spaces immensely great, and perhaps not less than our whole solar system.

CHAPTER II

CLASSIFICATION, NUMBER, AND DISTRIBUTION

CLUSTERS have been described and classified as loose, irregular, ragged, coarse, open, furrowed, poor, galactic, globular, open globular, rich, condensed, nebulous, with various qualifying adjectives for each class   All of these descriptions are necessarily based on superficial appearances and, because of the imperceptible gradation from one form to another, are little more than working conveniences.   Recently, however, the globular clusters shown on Harvard plates have been classified according to central concentration,[1] and both Trumpler[2] and I have proposed simple classifications of the brighter galactic clusters in terms of the spectral characteristics of their stars.

3. A Comparison of Galactic and Globular Clusters.—It is proposed to adopt in the present treatment only two main divisions—globular clusters and galactic clusters.[3]   Globular clusters may be "typical," like Messier 13, open, like Messier 4 and N. G. C. 3201, or elongated, like Messier 19.   Their principal characteristics are strong central concentration and richness in faint stars.   The galactic clusters are extremely varied; for example, Messier 11 is relatively rich, Messier 35 is irregular, the Pleiades and Messier 16 are nebulous, and Messier 103 and N. G. C. 1981 may be but accidental groupings. The so-called moving clusters are merely the brighter and nearer of the galactic types in which radial or transverse motions

---

[1] Shapley and Sawyer, H. B. 849, 1927.

[2] P. A. S. P., **37**, 307, 1925.

[3] The term "galactic cluster," suggested by Trumpler (P. A. S. P., **37**, 307, 1925) and others, is a natural name for the non-globular cluster, which is almost without exception near the galactic plane.   It replaces the term "open cluster," which has caused some confusion because of the open type of globular cluster.

have been measured. ·The broad category of moving clusters includes Praesepe and the Pleiades as well as Ursa Major, Scorpio-Centaurus, and similar systems (see Chapter VII, Section 35).

Some of the irregular galactic clusters resemble the small galactic star clouds; and such star clouds near the galactic circle are similar in many features to star clouds outside the Galaxy—that is, to stellar organizations like N. G. C. 6822 and the Magellanic Clouds, which lie well beyond the main body of galactic stars. These clouds, in turn, are typical of a considerable class of extra-galactic systems and apparently grade directly into the spiral nebulae and affiliated forms.

In future studies, especially of the Magellanic Clouds, we may find further examples of clusters in a transitional stage, between the richer galactic groups and the most open globular clusters. At present, however, there seems to be a rather sharp division which distinguishes the globular clusters as a special group of sidereal organization—a group limited to about 100 objects[4]—while the galactic ·clusters grade indefinitely into multiple stars in one direction and, in another, as indicated above, into small irregular star clouds.

Clear discrimination between galactic and globular clusters is also possible on the basis of distribution in the sky. The subject had been considered[5] with more or less care by Bailey, Bohlin, Hinks and Hardcastle, Melotte, and Perrine before it was taken up by the writer as a part of the systematic work on clusters. (The most conspicuous feature of the distribution is that galactic (open) clusters are almost exclusively in the Milky Way and distributed irregularly throughout all galactic longitudes, while the globular clusters are rather widely scattered in latitude but quite restricted in longitude) (The globu-

[4] The future resolution of external star clouds will probably disclose many apparently globular clusters, but because of their great distance it may never be possible to say definitely that these remote groups are not merely rich clusters of the galactic type or even poor clusters involved in strong nebulosity.

[5] See the general bibliography in Appendix C

lar clusters are, in fact, mostly in one half of the sky, as will be shown in subsequent diagrams.

The considerations outlined above justify the division of clusters into the two main groups. The subdivisions described below have been found of practical use in the study of clusters and are perhaps indicative also of the various stages of development or decay, though they are essentially descriptive and do not directly imply an evolutionary theory.

4. Classification of Galactic Clusters.—In the studies of galactic clusters at Harvard we have for some time followed a two-dimensional classification. One parameter is related to the apparent number and concentration of the stars and may be called "compactness"; the other depends on the distribution of spectral classes among the cluster members.

The classification based on appearance is intended to cover the whole range of galactic clusterings, from multiple stars to globular clusters. The subdivisions are as follows:

a. Field Irregularities.—That there are many deviations from random stellar distribution is obvious from star counts, or even from a casual inspection of photographic plates in nearly any region of the sky. Such "excess irregularities" have been referred to in Chapter I, where it is pointed out that neither dark nebulosity nor random distribution can account for the many faintly outlined non-uniformities in stellar fields. These incipient or vestigial star clouds vary in population from a few scattered members to vast indefinitely limited congregations of stars. There seems to be no immediate need of attempting to unravel or catalogue them; but the assignment of a classification letter to the field irregularity is recognition of its significance in stellar distribution.

b. Star Associations.—In this category fall wide-spread moving clusters, such as the Ursa Major group, and the peculiar stars of high and parallel velocities. The class will be recruited largely through studies of proper motion and radial velocity. It grades imperceptibly into the next class.

c. Very loose and irregular clusters, typified by the Hyades and Pleiades. The large cluster of bright stars around α Persei might be placed in this class or, better, perhaps, placed with the Orion nebula cluster in Class b. Class c corresponds, in general, with Bailey's D3 and with Melotte's IV.[6]

d. Loose Clusters.—Messier 21 and Messier 34 are examples of a class equivalent to Bailey's D2 and Melotte's III.

e, f, g. Compact Clusters.—These groups are equivalent to Bailey's D1 and Melotte's II. The division into three types is made on the basis of richness and concentration; examples are Messier 38, Messier 37, N. G. C. 2477. In the classification of clusters, the globular systems follow immediately after Class g. In fact, several of the most compact Class g galactic clusters appear more nearly like globular clusters than do the loosest globular clusters classified as such by criteria other than appearance.

In practice, the galactic clusters are generally taken to comprise only classes c to g. A number of the classes are illustrated in Plate II. A further consideration of the size and structure of the various systems is deferred to Chapter VII. The distribution among the classes of the 249 clusters listed in Appendix B is as follows:

Class	Number
c	20
d	85
e	67
f	47
g	30

The preceding classification is based on appearance and depends mainly on the population and distance of a group; it is independent of the spectra of the stars involved or of the stage of development of the cluster. I have found it convenient to divide galactic clusters also into two principal groups on the basis of the spectra or colors of the component stars—(1) the Pleiades type and (2) the Hyades type. Each includes members of classes c to g. In the Pleiades type, the stars, almost

[6] Bailey, II. A , 60, 200, 1908; Melotte, Mem. R. A. S., 60, 175, 1915.

without exception, lie along the "main branch" of a Russell diagram, with the earliest classes B or A; and in the Hyades type, yellow spectral classes occur with the same apparent brightness as the predominant A stars.

The tabulation of the brighter stars in the Pleiades and the Hyades on page 32 fully illustrates the difference between the two major types. More than 95 per cent of the galactic clusters for which spectral classes or colors have been determined fall in one or the other of these two types, which are about equally numerous. There are a few aberrant clusters. Messier 67, for instance, appears to be a variant of the Hyades type, in that blue giant stars are absent. Prominent examples of the Pleiades type are the double cluster in Perseus, Messier 36, and Messier 34; the Hyades type includes Messier 11, Messier 37, Praesepe, and the scattered cluster in Coma Berenices.

Trumpler has also proposed and used a classification of galactic clusters based on spectral composition.[7] For clusters that he has so far observed, he uses types 1a, 1b, 2a, and 2f, with provision for other types if found. Figure III, 9 shows the distribution of spectral classes among the four types. In Type 1 he includes clusters from which the giant branch (yellow and red stars) is entirely missing or in which there are so few scattered stars falling within their limits that it is uncertain whether they are physical members or background stars. This is the equivalent of my Pleiades type, where all the stars fall along the main sequence. Type 2, corresponding to my Hyades type, comprises "the clusters which show a marked crowding of stars along the giant branch although their number may still be small compared with that of the dwarf stars."[8] Remarks on the possible significance of the two main types of galactic clusters will appear in a later chapter.

[7] P. A. S. P., **37,** 307, 1925. A more recent study of galactic clusters is presented by Trumpler (L O B **14,** No. 420, 1930), where he gives a three-dimensional classification. The investigation includes statistical material on distances, magnitudes, dimensions, and spectra
[8] *Ibid.*

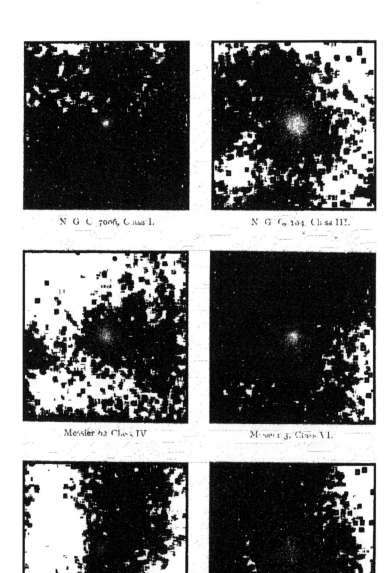

N G C 7006, Class I.

N G C 104, Class III.

Messier 62, Class IV.

Messier 3, Class VI.

Messier 19, Class VIII

Messier 4, Class IX.

PLATE I.—GLOBULAR STAR CLUSTERS.

5. Classification of Globular Clusters.—Notwithstanding, the general similarity of globular clusters in size, form, content, and absolute brightness, there are many deviations from the average. Clusters such as Messier 19 and ω Centauri are conspicuously elongated; Messier 62 is strikingly non-symmetrical; N. G. C. 4147 is deficient in giant stars; and, as intimated above, nearly one third of the globular systems are so loosely organized that from a casual examination, photographic or visual, we might place them with the galactic clusters and exclude them from their true affiliation. That such loosely built systems are of the globular class is made certain by long-exposure photographs which bring out thousands of faint stars such as have never been observed in even the richest of galactic clusters; and their attribution to the globular class is also often indicated by high galactic latitude and by the presence (as in Messier 4, Messier 72, N. G. C. 3201, N. G. C. 5053) of many cluster-type Cepheid variables.

Until recently, however, no systematic attempt has been made to classify the globular clusters beyond noting that some were variable-rich, some variable-poor; some open, some compact. I proposed a few years ago[9] that N. G. C. 7492 might be taken as a type of a rather distinct subdivision, called the "loose globular cluster," which would include those parenthetically mentioned in the preceding paragraph and N. G. C. 288, I. C. 4499, and N. G. C. 5466.

A detailed examination by Miss Sawyer and the writer of the globular clusters on good Bruce photographs, which are available in the Harvard collection for practically all the 103 systems now listed as globular, shows that many intermediate forms exist between the loosest and most concentrated clusters. Instead of classing the clusters, therefore, in the two or three broad and obvious categories, such as compact, medium, loose, we arrange them in finer subdivisions, in a series of grades on the basis of central concentration. In Table II, I is given a list of globular clusters chosen as typical of the 12 subdivisions.

[9] Mt. W Contr 161, 1918

Photographs of six of the representative types are reproduced in Plate I. For the classification of individual clusters, reference may be made to Appendix A.

TABLE II, I.—TYPICAL GLOBULAR CLUSTERS OF THE TWELVE CLASSES

Class	N. G. C.	R A (1900)		Dec (1900)		Pg Mag	Class	N. G C	R A (1900)		Dec (1900)		Pg Mag
		h	m	°	′				h	m	°	′	
I	2808	9	10 0	−64	27	5 7	VII	6656	18	30 3	−23	59	3 6
II	7089	21	28 3	− 1	16	5 0	VIII	6402	17	32 4	− 3	11	7 4
III	104	0	19 6	−72	38	3	IX	6218	16	42 0	− 1	46	6 0
IV	1866	5	13 3	−65	35	8 0	X	288	0	47 8	−27	8	7 2
V	7099	21	34 7	−23	38	6 4	XI	6809	19	33 7	−31	10	4 4
VI	6752	19	2 0	−60	8	4 6	XII	7492	23	3 1	−16	10	10 8

It is not likely that detailed counts of stars will always arrange the clusters in the order shown by the concentration class. The numerical concentration, determined from counts, certainly varies with stars of different magnitudes, and because of crowding and Eberhard effect it will always be of doubtful value except for the brighter stars. Our estimated concentrations are also slightly influenced by the quality of the plates and the total brightness and angular diameters of the clusters, but we believe that these factors are not of such consequence that they detract appreciably from the value of the classification. Class I represents the highest concentration toward the center, and Class XII the least. The distribution among the various classes and the mean photographic magnitude for each class are as follows:

Class	Magnitude	Number	Class	Magnitude	Number
I	8 85	4	VII	7 90	8
II	7 80	7	VIII	7 85	10
III	6 76	7	IX	8 84	10
IV	9 01	12	X	8 88	9
V	7 88	12	XI	9.54	9
VI	8.91	11	XII	9 58	4

The present classification of globular clusters is essentially a description of apparent central concentration. It is interesting, therefore, to note that there is no correlation of class with integrated photographic magnitude as determined from Harvard plates of small scale. The distribution of magnitude among the various classes is shown in the scatter diagram in Figure II, 1, where also the mean magnitude for a given class is plotted as a cross, and the average class for each interval of one magnitude is plotted as a circle.

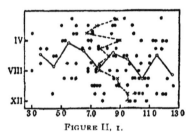

FIGURE II, 1.

The scatter diagram of classes of globular clusters (ordinates) and integrated photographic magnitudes Circles and crosses indicate means

The galactic distribution of the most concentrated clusters does not differ measurably from that for the least concentrated. We have, for example:

Classes I to VI   Mean longitude 265°   Mean latitude ± 23°
Classes VII to XII   Mean longitude 263°   Mean latitude ± 21°

For these computations the globular clusters in the Magellanic Clouds are excluded.

To maintain homogeneity, the classifications of globular clusters were all made on plates with the scale of 1 mm = 1′. Superposed stars have occasionally interfered somewhat with the assignment of the class, especially for N. G. C. 4147, 6284, 6453, 6553, 6569, 6624. A few peculiarities were noted that are not completely taken care of by our classification based on central condensation alone. For instance, the bright cluster ω Centauri is peculiar in what appears to be a remarkable uniformity in the magnitudes of the brighter stars. Clusters somewhat similar to ω Centauri in this respect are N. G. C.

5272 (Messier 3), 5927, 6273, 6656 (Messier 22). These clusters also resemble each other in their moderate concentration (classes VI to VIII), and two of them, ω Centauri and Messier 3, are the richest of all in variable stars. It should be noted, however, that the clusters with many variable stars are scattered throughout all classes.

In conclusion, we observe that the classes of globular clusters are probably indicators of developmental age. They should prove increasingly useful in studies of linear diameters, motions, luminosity curves, and the deeper problems of the origin and life history of stellar clusters.

6. The Number of Clusters.—Since the time of the Herschels, very few globular clusters have been discovered, notwithstanding the considerable increase in telescopic power and the great increase in the known number of stars and nebulae. Every recognized globular cluster except one bears a number from the New General Catalogue of Dreyer, an indication that in spite of great distance and the faintness of the individual stars, all of these objects were known prior to 1880. They were known, indeed, before 1864, the date of Sir John Herschel's General Catalogue, and all but a few were catalogued more than 90 years ago in the earlier Herschelian lists. This early completion of the discovery of the globular clusters led Bailey[10] to suggest that the limit of the region occupied by these systems had been reached, a suggestion that appears to be supported by subsequent work. Thousands of new nebulae and millions of stars have been added by modern telescopes and photographic plates, but the essentially complete listing of globular clusters antedated photography.

There is, however, a vast difference between cataloguing an object and recognizing its true character. Many of the entries given in the N. G. C. as globular clusters have proved to be something else, generally galactic groups or extra-galactic nebulae; and 34 of the globular clusters now recognized were

[10] H. A., 76, 43, 1915.

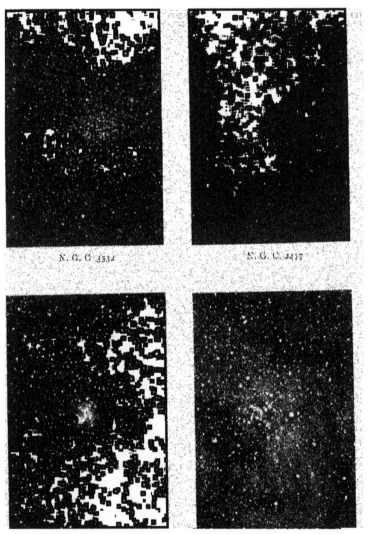

N. G. C. 3534   N. G. C. 3477

Messier 16   Messier 7

PLATE II.—GALACTIC STAR CLUSTERS.

not described as such in the New General Catalogue. The large photographic telescopes have been of service in recent years in examining many faint and doubtful N. G. C. objects, and an occasional addition to the list of globular clusters has resulted. A number of remote groups remain doubtful, however, even after some of them have been tested with large reflectors. Chief among the doubtful objects are the following:[11]

N. G. C.	Radec*	Galactic	Distance†	Note
1651	0438 − 71	249 − 36		Rejected, see H. C. 271, 1925
5946	1528 − 50	295 + 04	32 2	A small, poor, loose cluster in a rich region
6352	1718 − 48	308 − 07	19.7	A comparatively large cluster of very faint stars, on the edge of the Milky Way
6426	1740 + 03	356 + 15	37 1	Very faint and poor; suggestion of a background on Mount Wilson plates
6535	1759 − ∞	354 + 10	26 7	A small cluster, on the edge of a rich region, with few stars
6539	1759 − 08	348 + 06	38 7	A very faint cluster in a large obscured area
6712	1848 − 09	353 − 06	26 2	A little, irregular knot of stars in a rich cloud
6760	1906 + 01	003 − 05	28.6	A faint, sparse, loose cluster in a rich region

* The approximate positions for 1900 in equatorial coordinates are conveniently contracted for tabulation into the form here given, the first four figures give the hours and minutes of right ascension, and the sign and subsequent figures indicate the declination in degrees (and may be extended to minutes, if desired).

† The distance in kiloparsecs is estimated on the assumption that the clusters are globular.

Sixty-six globular clusters were included in the catalogue of bright clusters and nebulae compiled in 1908 by Bailey.[12] To this number Hinks[13] added 41 from Dreyer's catalogue, though several of the additions are not now accepted. Bohlin[14] considered about 75 objects truly globular. Miss Clerke[15] stated that approximately 500 clusters of all kinds were known, 120 having globular forms. In later publications, Bailey[16] placed the total number of globular clusters at 76, and the total

[11] Sawyer and Shapley, H. B. 848, 1927.
[12] H. A , 60, 199, 1908.
[13] M. N. R. A. S , 71, 697, 1911.
[14] Swedish Acad., 43, No. 10, 1909.
[15] System of the Stars, 227, 1905.
[16] H. A , 76, 43, 1915.

of all kinds at nearly 700.   This high total includes many chance aggregations and a number of minor condensations in the Milky Way.

The data on which the above estimates are based lack homogeneity.   Melotte's catalogue,[17] however, which was made from the Franklin Adams plates and contains 83 globular and 162 open clusters, constitutes a fairly homogeneous list of all clusters with diameters greater than one minute of arc and brighter than the sixteenth or seventeenth photographic magnitude. His list of galactic clusters is revised and extended in Appendix B, which contains 249 entries.   From his list of globular clusters a few objects have been dropped; others have been added, mainly as a result of my work and Hubble's with the 60- and 100-inch reflectors at Mount Wilson.   As stated above, the total number, those accepted for the tables in Appendix A and 10 globular clusters in the Magellanic Clouds, is 103.   Further additions will probably be slow and will come through the closer analysis of small groups in and near the Milky Way star clouds and the detailed examination of small nebulous objects that are now assumed, without very good reason, to be elliptical extra-galactic nebulae.

An example of a recent addition is the observation at the Lowell Observatory, verified at Mount Wilson, that the object N. G. C. 2419, described in the New General Catalogue as "pB, pL, lE 90°, vgbM, *7.8 267°, 4' dist," and not listed by Melotte, is, in fact, a remote globular cluster in the part of the sky that is otherwise devoid of these systems.[18]

It may be well to recognize at this point that if the spheroidal extra-galactic nebulae are actually stellar throughout, perhaps many of them are essentially globular star clusters, probably at much greater distances than the objects here studied and of a greatly different order of dimensions.   In the list of faint nebulae, shown on Bruce plates made at Arequipa and on similar photographs, there are large numbers of very small

[17] Mem. R. A. S., **60**, 175, 1915.
[18] Shapley, H B. 776, 1922.

unresolved circular images that are listed hopelessly as nebulae of the spiral family. Occasionally, one of these may be a very remote (but ordinary) globular cluster. A comparative study of the distribution of light throughout the images can, however, give us some indication of their nature; their distribution with respect to other extra-galactic objects will probably show most of them to be sidereal systems of a higher order than globular clusters.

7. The Apparent Distribution of Galactic Clusters.—The new catalogue of galactic clusters has been prepared on the

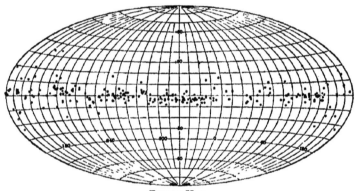

FIGURE II, 2.

Distribution of galactic clusters in galactic coordinates. Cluster classes are indicated as follows c, O, d, ⊕, e, ◐, f, ◑, g, ●.

basis of a fairly uniform series of Harvard photographs(Appendix B). The plates studied were made with the 8-inch Bache and Draper and the 10- and 16-inch Metcalf telescopes. Scores of loose groups of a few stars, which appear in the N. G. C. and in other lists, are omitted from the new catalogue. At the same time, 21 new clusters, never before listed, are now included. The total number for which we consider the distribution is 249.

Figure II, 2 shows the distribution of galactic clusters in galactic coordinates. The high concentration in low latitudes is immediately evident. The galactic clusters with latitudes greater than ± 15° are given in Table II, II.

TABLE II, II.—GALACTIC CLUSTERS WITH LATITUDE GREATER THAN ±15°

N. G. C.	Galactic		Remarks
	Longitude	Latitude	
	°	°	
188	90	+23	
752	105	−23	
Melotte 22	134	−22	Pleiades
Melotte 25	147	−23	Hyades
2243	206	−16	
2281	142	+18	
2420	166	+21	
2548	196	+17	
2632	174	+34	Praesepe
2682	184	+33	Messier 67
Melotte 111	200	+85	Coma Berenices
I 4665	358	+16	

The high latitudes for Coma, Praesepe, the Pleiades, the Hyades are not remarkable, because the parallaxes are relatively large and the linear distances from the galactic plane accordingly are relatively small. A further discussion of the distribution of galactic clusters is to be found in Section 70 below.

An interesting and significant result of the special surveys that have been made of objects of doubtful class is the evidence that every faint little-condensed cluster in galactic latitude higher than 15° or 20° is really globular, although for many of them short exposures and visual observations had originally recorded few stars. Long exposures, however, invariably bring out the globular nature of the objects, with the possible exception of N. G. C. 5053, noted below. All the similar faint objects along the galactic equator remain open groups, with no condensed background of faint stars appearing on long exposures.

8. N. G. C. 5053 and N. G. C. 2477.—The most conspicuous apparent deviation from the rule of low latitudes for galactic clusters is for the faint object N. G. C. 5053. Its latitude is +77°. Baade[19] first noticed that this object, described as

[19] Baade, Hamb Mitt , 5, No 16, 1922

"Cl, vF, pL, iR, vgbM, st 15" in the N. G. C., appears to be an open cluster of faint stars. His longest exposures with the Bergedorf reflector, however, left the matter indeterminate. At my request Dr. Hubble made an exposure of 90 minutes with the 100-inch reflector. From this plate the cluster does not seem to be globular; that is, no concentrated background of faint stars appears at the center—the total number of the stars appears to be a few hundred.

More recently Baade[20] has found in N. G. C. 5053 eight variables of the cluster type, which certainly cannot be considered random members of the foreground. No galactic cluster has variables of this sort; they are common in globular systems. Baade also finds the general luminosity curve of N. G. C. 5053 similar to those I have obtained for globular clusters. Its population, however, is only a fourth that of a typical globular system such as Messier 3. On three counts, therefore, we call N. G. C. 5053 a globular cluster—its position, its variables, and the frequency of absolute magnitudes.

In superficial appearance N. G. C. 2477, galactic latitude $-5°$, is the richest of galactic clusters; or perhaps it is the loosest of globular clusters. No variables have been found within it, but the examination has not been exhaustive. Miss Sawyer has determined the general luminosity curve to magnitude 16.8 for a circle with radius 14'.0. The correction for the foreground and background stars (232 in all) was made on the basis of star counts in adjacent fields. The results are as follows:

Limiting Magnitudes	Cluster Stars	Total Stars	Limiting Magnitudes	Cluster Stars	Total Stars
]10 0	0	0	13 5–14 0	134	162
10 0–10 5	0	3	14 0–14 5	121	153
10 5–11 0	2	4	14 5–15 0	115	158
11 0–11 5	3	4	15 0–15 5	74	109
11 5–12 0	3	10	15 5–16 0	51	76
12 0–12 5	17	30	16.0–16 5	58	76
12 5–13 0	79	89	[16 5	30	35
13 0–13 5	117	126			
Total				804	1,036

[20] *Ibid*, 6, No 29, 1927

For the present we leave N. G. C. 2477 in the list of galactic clusters, awaiting further photometric and spectroscopic analysis with larger telescopes.

9. The Apparent Distribution of Globular Clusters.— In a form comparable to that of the preceding diagram, the

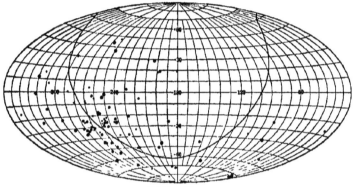

FIGURE II, 3

Distribution of globular clusters in equatorial coordinates   Small dots indicate more distant clusters.   The galactic circle is shown by a heavy line.   The clusters in the Magellanic Clouds have been omitted

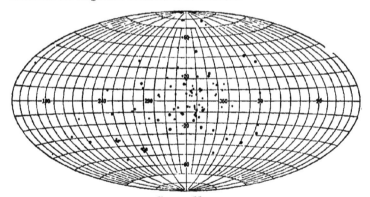

FIGURE II, 4.

Distribution of globular clusters in galactic coordinates.

distribution of globular clusters is shown in Figures II, 3 and II, 4.   The equatorial and galactic coordinates of each cluster are given in Appendix A, and a discussion of their distribution in space appears in later chapters.   The remarkable contrast in the galactic affiliation of the galactic and globular clusters

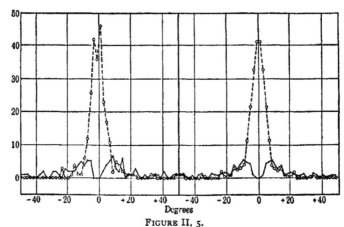

FIGURE II, 5.

Numbers of galactic clusters (circles) and globular clusters (dots) for two-degree intervals in galactic latitude   Left, hemispheres separately; right, means.

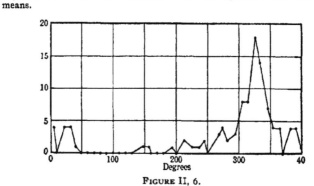

FIGURE II, 6.

The frequency distribution of globular clusters in galactic longitude   Ordinates are numbers of clusters; abscissae, degrees of longitude.

is evident from the diagrams and is further illustrated in Figure II, 5.   The remoteness of the globular clusters suggests that obstruction by dark nebulosity in the Milky Way can easily influence their observed scarcity near the galactic equator without affecting at the same time the distribution of the galactic clusters.

In Figure II, 6 is shown the distribution of globular clusters in galactic longitude. From this diagram and the preceding figure we find that the center of the system of globular clusters lies in the direction of galactic longitude 327°, galactic latitude 0°. The probable error of this determination is about 1°. The corresponding equatorial coordinates are right ascension 17^h 28^m, declination $-29°$.

10. Clusters in or near Obstructing Nebulosity.—The large groups of bright B stars in Orion, Scorpio, and elsewhere are associated with important bright and dark nebulosity. The Pleiades nebulosity is well known. A number of nebulous galactic clusters, such as Messier 8, are on record, and clusters of this sort also appear in the Magellanic Clouds. It is doubtful, however, if much would be known of the nebulosity in these galactic clusters if their distances were ten times as great. By implication, therefore, nebulosity in galactic clusters may be more common than appears from general inspection.

There are a few individual globular clusters, N. G. C. 4372, N. G. C. 6144, and N. G. C. 6569, that are in or near recognized dark or luminous nebulae. Of these, the first appears to be dimmed by one of the long dark streamers from the Coal Sack; the second is at the edge of the heavy ρ Ophiuchi nebulosity. N. G. C. 6569 is in a rich star field in Sagittarius but may also be involved in wisps of obscuring nebulosity.

To what extent the magnitudes and colors in galactic and globular clusters are directly affected by associated nebulosity is not as yet determined. For individual nebulous stars, Seares and Hubble have found a color effect and presumably a corresponding deficiency in apparent brightness.

CHAPTER III

ON THE SPECTRAL COMPOSITION OF CLUSTERS

THE variety of spectral classes in galactic clusters and the differences in composition from system to system have long been known; the Pleiades and the Hyades, for example, exhibit wide diversity and contrast. Also, among globular clusters some have been noted as relatively yellow, others white; but their contrasts are less marked. The first definite observation of color in a globular cluster was made by Barnard,[1] who compared photographic magnitudes with visual and photovisual magnitudes for a number of stars in Messier 13, the Hercules cluster. Further determinations of colors and spectra in globular clusters[2] have been based primarily on work with the Mount Wilson photographs, which will be described in Section 12.

11. **Integrated Spectra of Globular Clusters.**—Integrated spectra were determined for several clusters many years ago at the Lick, Mount Wilson, and Lowell observatories, mainly by Fath[3] and Slipher,[4] who found spectra of composite character, principally classes F and G. The Henry Draper Catalogue records without close classification the integrated spectra of numerous clusters. At my request Miss Cannon has recently examined closely all available Harvard photographs showing spectra of globular clusters.[5] In classifying more than 40 integrated spectra she has greatly increased our knowledge of the average effective composition of such systems.

[1] Ap. J., 12, 176, 1900; 29, 72, 1909; 40, 173, 1914.
[2] See Appendix D for special bibliography on spectra.
[3] L. O. B., 5, 74, 1908; Mt. W. Contr. 49, 1911.
[4] Pop. Astr., 25, 36, 1917; 26, 8, 1918.
[5] H. B. 868, 1929.

The results are briefly summarized in Table III, I, which gives the N. G. C. designation, the cluster class, and the spectrum. For some clusters, such as N. G. C. 4147 and N. G. C. 6624, the observed spectrum may be largely due to one or two stars, and they, of course, may be foreground objects. A frequency diagram of the classes is given in Figure III, 1; classes F, G, and K of Table III, I are taken as Fo, Go, and Ko. The class of spectrum does not appear to be closely related to cluster

TABLE III, I.—SPECTRA OF GLOBULAR CLUSTERS

N G. C.	Cluster Class	Spectral Class	N G. C.	Cluster Class	Spectral Class
104	III	G5	6293	IV	G5
362	III	G5	6304	VI	K:
1261	II	G	6316	III	G5
1851	II	Go	6333	VIII	K·
1866	IV	F8	6341	IV	G5:
1904	V	F8	6356	II	Ko
2808	I	Ko	6388	III	K
4147	IX	A7:	6397	IX	G:
4590	X	Note	6441	III	Ko
5024	V	Note	6541	III	G
5272	VI	G	6624	VI	Mo
5286	V	Go	6626	IV	G5
5824	I	F8	6637	V	K2
5904	V	G:	6652	VI	K5
5986	VII	F8	6715	III	F8
6093	II	Ko	6723	VII	G5:
6121	IX	F	6752	VI	Go
6205	V	Go	6864	I	Go
6229	VII	Note	6934	VIII	Go
6254	VII	Note	7078	IV	F
6266	IV	Ko	7089	II	F5
6273	VIII	G5	7099	V	F8
6284	IX	F:			

NOTES TO TABLE III, I

N. G C. 4590. Dark lines are seen in the violet.
    5024. Dark lines Hδ, H, and K faintly seen.
    6229. Several dark lines are seen, apparently including H and K
    6254   Dark lines in the violet appear to be H, K, and Hγ

class, apparent magnitude, or any other significant property of the clusters. We have here only an indication of the probable diversity in predominant type of the stars that are effective in producing the integrated spectrum.

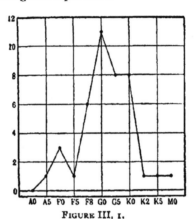

FIGURE III, I.

Integrated spectra of globular clusters. Coordinates are numbers of clusters and spectral classes.

12. Stellar Types in Globular Clusters.—From integrated spectra and colors we pass to the types of individual stars in globular clusters, noting some similarities to the distribution in the Galaxy but also some important differences. The discussion of variable stars and their problems is left to the next chapter.

a. Common Spectral Classes.—The color indices in various globular clusters, especially Messier 13[6] and Messier 3[7], show a normal range from about −0.3 to +1.6.[8] This indicates, no doubt, the presence of all spectral classes from B to M; and normal spectra from A to G have, in fact, been directly observed.

The spectra of about 50 individual stars in Messier 13 and other bright globular clusters have been classified by Pease and

[6] Shapley, Mt. W. Contr. 116, 1915.
[7] *Ibid.*, Mt. W. Contr. 176, 1919.
[8] *Ibid.*, Mt. W. Comm. 44, 1917.

Sanford[9] on plates made at Mount Wilson, but the details of the work have never been published.   In Messier 13 Pease finds the classes of 39 stars distributed as follows (classifications by Adams):

Class	Number
A0	3
A5	11
F0	11
F5	11
G0	3

The relative frequency of the various classes probably differs somewhat from one cluster to the next, as indicated by the diversity of the integrated spectra, but sufficient observations are not yet available to decide this rather important matter. The most frequent color index in Messier 13 (outside the unresolved center where the heavy stellar concentration and the Eberhard effect have prevented satisfactory work) is about +0.70, and the average color index is +0.55, corresponding to the spectrum gF4.   Seventeen per cent of the stars brighter than the working limit (photovisual magnitude 15.5) have negative color indices, indicating spectral classes earlier than A.   In Messier 3, on the other hand, there is a smaller proportion of negative color indices (4.7 per cent down to photovisual magnitude 17.00) but an excess of Class A stars of about the magnitude and color of the cluster-type variables.   It should be noted, however, that an error in the zero point of either photovisual or photographic magnitudes would shift the spectral frequency curve bodily.   Such error may exist and may not be inappreciable.

✓ b. The Color-magnitude Arrays.—The distributions of color indices and photovisual magnitudes are shown in Tables III, II and III, III for Messier 13 and Messier 3, respectively.[10]

[9] Pease, Mt. W. Ann. Rep., 9, 219, 1913; 10, 268, 1914.   Sanford, ibid., 14, 212, 1918; Pop. Astr., 27, 99, 1919.
[10] Ten Bruggencate has plotted the coordinates (my magnitudes and colors) for the individual stars used in making these color-magnitude summaries, and has sought particular significance in the details of the resulting "Farbenhelligkeits-diagramme" (Sternhaufen, Berlin, 1927).   I think that there is little gain and

TABLE III, II—COLOR-MAGNITUDE ARRAY FOR MESSIER 13

Limits of Photovisual Magnitude	Color Class										All Colors
	bo to b5	b5 to ao	ao to a5	a5 to fo	fo to f5	f5 to go	go to g5	g5 to ko	ko to k5	k5 to mo	
12 00–12 19								1	3	2	6
12 20–12 39									1	1	2
12 40–12 59								3	2		5
12 60–12 79				1	1		1	2			5
12 80–12 99							4	2			6
13 00–13 19	1					1	8	3			13
13 20–13 39						1	4	2	1		8
13 40–13 59						1	6	1			8
13 60–13 79						2	3	1		1	7
13 80–13 99		3		2	6	14	6	2			33
14 00–14 19					1	9	3	2			15
14 20–14 39			1			10	13	4			28
14 40–14 59		1	1	7	3	24	16	4			56
14 60–14 79			6	2	1	6	12	1	1		29
14 80–14 99		5	3		2	5	8	2			25
15 00–15 19		24	9	3	3	19	10	2			70
15 20–15 39	10	21	7	5	12	28	15				98
15 40–15 59	4	11	6	4	11	11	5	1	1		54
15 60–15 79	1	5	3	3	10	4	1				27
Total	16	70	36	27	50	135	115	33	9	4	495

The tabulated quantities are numbers of stars. For Messier 13, no stars within 2' of arc of the center are included; for Messier 3, the limits are 2'.0 to 11'.3. The perils arising from the Eberhard effect are thus largely avoided.

Similar arrays are given in greater detail for these two clusters, for Messier 68, and for the galactic clusters Messier 67 and Messier 11 in my special discussions of the individual systems (see Appendix C). In all arrays the last tabulated line or two are near the practical fainter limit for magnitude work and may be deficient in numbers and inaccurate. The color classes, b, a, f, . . . , corresponding to the color index intervals[11] —0.4 to 0.0, 0.0 to +0.4, +0.4 to +0.8 . . . , are nearly

some danger in detailed subdividing of the observational material. Actual experience with the photometric measures in globular clusters leads to a belief that the group values presented in my color-magnitude arrays go as far as is justifiable in subdivision.

[11] Seares, Mt. W Comm. 16, 1915.

TABLE III, III.—COLOR-MAGNITUDE ARRAY FOR MESSIER 3

Limits of Photovisual Magnitude	Color Class										All Colors
	bo to b5	b5 to ao	ao to a5	a5 to fo	fo to f5	f5 to go	go to g5	g5 to ko	ko to k5	>k5	
]12 00	.			1				1	1		3
12.00–12 19											
12.20–12 39										1	1
12.40–12 59							.		1	1	2
12.60–12 79					1		1			2	4
12 80–12 99					1			4			5
13.00–13 19								5	2		7
13 20–13 39								1	2		3
13.40–13 59					1			4	1		6
13 60–13 79							4	3			7
13.80–13 99					1		7	6			14
14.00–14 19	.		1			2	3	5	1		11
14.20–14 39						1	3	3			7
14 40–14 59		1		1	3	16	6				27
14.60–14 79	1			1	1	11	11				27
14 80–14 99				1	1	3	7	4			16
15 00–15 19			1	2		7	17	3			30
15 20–15 39		1	4	4	15	19	13	1			57
15.40–15 59		9	23	12	15	26	1				86
15.60–15 79		8	6	4	2	28	8	2			58
15.80–15 99		2	6		7	17	5				37
16 00–16 19	2	2	2		12	14	6				38
16 20–16 39	1	1		4	16	13	1	1			37
16 40–16 59		1		9	21	19					50
16 60–16 79			2	20	26	11	1				60
16.80–16 99		2	4	24	38	1					69
Total	4	27	49	83	158	191	92	42	12	4	662

analogous to the spectral classes B, A, F, . . . This analogy is close because the cluster stars under study are all giants. The agreement would not be so satisfactory for dwarfs.

The general similarity of globular clusters, especially in the color-magnitude relation for giant and supergiant stars, is nicely illustrated in Table III, IV, where the brightest stars in two of the nearby globular clusters are compared with the brightest stars in the faintest and most remote globular cluster known. N. G. C. 7006 is five times as distant as Messier 3 and Messier 13. The tabulation shows that the brighter stars in all three systems have about the same average color

TABLE III, IV.—COMPARISON OF N. G C. 7006 WITH MESSIER 3 AND MESSIER 13

N. G. C. 7006			Messier 3			Messier 13		
Mean Pv Mag	Number of Stars	Mean Color Index	Mean Pv Mag.	Number of Stars	Mean Color Index	Mean Pv Mag	Number of Stars	Mean Color Index
15 56	5	+1 37	12 59	7	+1 30	12 11	6	+1 31
16 02	6	+1 14	12 90	5	+1 18	12 47	7	+1 14
16 41	7	+1 04	13 10	7	+1 61	12 72	5	+0 94
16 55	6	+1 16	13 43	9	+1 12	12 87	6	+0 82
16 82	7	+0 06	13 70	7	+0 99	13 05	6	+0 92
17 08	7	+0 95	(13 90)	(13)	(+0 96)	13 14	6	+0 93
Mean 16 46	38	+1 09	13 17	35	+1 15	12 72	36	+1 02

TABLE III, V *—COLOR-MAGNITUDE ARRAY FOR MESSIER 22

Limits of Photovisual Magnitude	Color Class													All Colors
	<bo	bo to b5	b5 to ao	ao to a5	a5 to fo	fo to f5	f5 to go	go to g5	g5 to ko	ko to k5	k5 to mo	mo to m5	>m5	
10 20-10.39													1	1
10 40-10.59														
10 60-10 79													1	1
10.80-10 99						1							3	4
11 00-11.19									1		3		3	7
11 20-11.39											3	6		9
11 40-11 59									2	2	1			5
11.60-11 79								1	2	1	1			5
11 80-11 99						1			2	5	1			9
12 00-12 19									3	1	1			5
12 20-12 39							2	2	8	2				14
12.40-12 59	.					1	3	5	2	2				13
12 60-12 79					1	1	6	15	2					25
12 80-12 99		.				1	3	4	3					11
13 00-13 19					1	1	4	3	1					10
13 20-13 39	.				1	6	9	3	1					20
13 40-13 59		1			1	8	6	6	1					23
13 60-13 79	.				1	12	28	16						57
13 80-13 99			1	1	4	34	59	7						106
14 00-14 19			2	1	17	39	12							71
14 20-14 39	1	1	11	40	61	39	3							156
14 40-14.59	.	1	19	26	20	2								68
14 60-14 79			3											3
Totals	1	3	36	68	105	138	123	57	45	17	10	12	8	623

* See Harvard Bulletin 874, 1930, and Chapter XIV below

and the same progression of color with magnitude.   A similar
result is found in all globular clusters tested, though some, such
as N. G. C. 4147 and N. G. C. 5053, are less populous in giant
stars. Occasionally, there are abnormally bright blue stars,
as in Messier 13, but even these are faint absolutely, compared
with some of the galactic B stars.

In the array for Messier 22 in Table III, V the small dispersion of magnitude within one color class is conspicuous.

The color-magnitude arrays establish the fact that in the
condensed clusters, as well as in some loose galactic groups, the
average color is redder the higher the visual brightness.
The result naturally bears on current consideration of the evolution of stars. The arrays call particular attention to the frequency
of the high-luminosity red stars, similar to Antares and Betel-
geuse, presumably of great mass.   The apparently universal
occurrence of red and yellow supergiants in globular clusters
argues for exceedingly slow development and consequently for
subatomic sources of stellar energy.   The presence of a single
supergiant, such as Antares, in a cluster might be ignored in
theoretical work—it might be treated as an abnormal occurrence.   But we find that in nearly every large assemblage of
stars there are reddish low-density objects which contraction
would have transformed in a few thousand years, but which
appear nevertheless to be untouched by age.   The question is
again taken up in Chapter XIV.

13. Distribution of Colors throughout Globular Clusters.
The relation of the colors of the brighter stars to their
positions in the clusters, although very important, is difficult
to determine satisfactorily, because the Eberhard effect is
certain to produce a spurious reddening in the crowded central
regions.[12]   Nevertheless, the conspicuous centralization of
red stars in the galactic cluster M 67, discussed later in Section
34, encourages an examination of the phenomenon in clusters
of the globular type.

[12] Shapley, Mt. W. Contr. 116, 58, 1915.

Observations that bear on the matter are contained in Figures III, 2, III, 3, and III, 4. The first two show the mean color indices (stars of all magnitudes) plotted against distance from

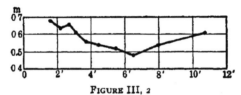

FIGURE III, 2
Change in Messier 3 of color index (ordinates) with distance from the center.

the center for Messier 3 and Messier 13. The plots are based on my observations contained in Mount Wilson Contributions

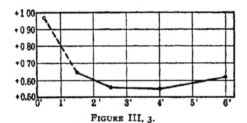

FIGURE III, 3.
Change in Messier 13 of color index (ordinates) with distance from the center.

116 and 176. The high point in Figure III, 3, connected with the next highest point by a broken line, is undoubtedly raised

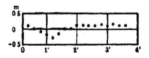

FIGURE III, 4.
Change in Messier 13 of color index (ordinates, with arbitrary zero) with distance from the center.

by the Eberhard effect. Neither cluster shows any very definite change in mean color with distance from the center, but stars fainter than absolute magnitude $+1$ are only slight factors in producing these results.

Figure III, 4 shows results derived by Hogg[13] from an investigation of the integrated light of Messier 13. The changes in color from the center to the edge are seen to be negligible. It appears, therefore, as far as the present data go, that the colors of the giant stars in globular clusters are not related to the radial distance from the center of the cluster.

√ **14. Types of Stars in Galactic Clusters.**—In the early work on stellar spectra it was found that the Pleiades are devoid of bright yellow stars, the most luminous members being of Class B. From the neighboring Hyades cluster, on the other hand, the B stars are completely absent; the bluest stars are of Class A, and intermingled with them are a few equally bright K stars. The visual magnitudes and the spectral classes of the ten brightest stars in each system are as follows:

Pleiades			Hyades		
Alcyone	2 06	B5e	θ_2 Tau	3 62	A5
Atlas	3 80	B8	ϵ Tau	3 63	K0
Electra	3 81	B5	γ Tau	3 86	G0
Maia	4 02	B5	δ Tau	3 93	K0
Merope	4 25	B5e	θ_1 Tau	4 04	K0
Taygeta	4 37	B5	Br 601	4 24	A0
Boss 879	5 18	B8e	Br 639	4 30	A5
Boss 851	5 43	B5	κ_1 Tau	4 36	A3
Boss 872	5 51	B8	v Tau	4 40	A5
Boss 861	5 85	B8	Br 605	4 60	A0

As long ago as 1897, Mrs. Fleming examined the spectra of individual stars in a number of loose galactic clusters, mainly southern. She noted not only that each cluster contains more than one spectral class but also that some clusters show a preponderance of blue stars, others of yellow or red stars,[14] thus foreshadowing the present general subdivision into Pleiades and Hyades types.

[13] H. B. 870, 1929.
[14] Pickering, H. A., 26, 1891, see Table III, VI.

Adams and van Maanen in 1913 noted the commonness of early B stars in the double cluster in Perseus,[15] and some unpublished colors and spectra that I obtained at Mount Wilson in 1919 showed that this system follows the Pleiades model in spectral distribution.

15. Spectra in Individual Galactic Clusters.—For a number of the largest galactic clusters there are fragmentary data in the Henry Draper Catalogue, the spectra of a few of the brighter stars having been classified in the routine course of the general program. To the extent of their value, these data have been used by various investigators, especially Doig[16] and Raab.[17]

The numerous investigations of the colors and magnitudes of galactic clusters are listed in Appendix C and separately tabulated in the special bibliography in Appendix D. Representative papers, giving the colors of individual stars, are those by Hertzsprung[18] and by Seares[19] for N. G. C. 1647, by von Zeipel and Lindgren[20] for Messier 37, by the writer and Miss Richmond[21] and by Graff[22] for faint stars in the Pleiades. All these studies emphasize the diverse spectral structure of galactic clusters. The spectral composition of the individual clusters has been discussed by Doig,[23] ten Bruggencate,[24] and Raab[25] and is in the process of exhaustive analysis by Trumpler[26] at the Lick Observatory. Some of these systems are described in Subsection b, and Trumpler's results are discussed in Subsection c. The Harvard material is taken up first.

[15] A. J., **27**, 187, 1913.
[16] J. B. A. A., **35**, 201, 1925.
[17] Lund Medd , Ser. 2, 28, 1922.
[18] Mt. W. Contr. 100, 1915.
[19] *Ibid.*, 102, 1915.
[20] Proc. Swedish Acad., **21**, No. 16, 1921.
[21] Mt. W. Contr. 218, 1921.
[22] Hamburg Abh., **2**, 3, 1920.
[23] J. B. A. A., **35**, 201, 1925.
[24] Seeliger Festschrift, p. 50, 1924.
[25] Lund Medd., Ser. 2, 28, 1922.
[26] P. A. S. P., **37**, 307, 1925.

a. Harvard Studies.—Mrs. Fleming[14] tabulated the spectra for seven galactic clusters: the Pleiades, Praesepe, Coma Berenices, I. C. 2602, N. G. C. 3532, Messier 6, and Messier 7. The stars for which spectra are classified are distributed over a larger area than that actually covered by the clusters, and foreground stars, of course, cannot generally be differentiated and excluded; very faint stars, barely distinguishable on the plates, are included. Some results are summarized for Class A in Table III, VI.

TABLE III, VI.—PERCENTAGE OF CLASS A STARS IN SEVEN GALACTIC CLUSTERS

Cluster	Stars Classified	Stars in H. D. C.	Per cent Class A
Pleiades	91	*	65
Praesepe	90	*	31
I. C. 2602	64	58	77
N. G C. 3532	204	135	93
Coma	117	*	15
Messier 6	91	75	75
Messier 7	346	177	78

* Limits indefinite.

Although the classification was crude, and for some stars quite uncertain, the existence is clearly shown of a spectral distribution in Praesepe and Coma different from that in the other five clusters. It is the same distinction that is now recognized in the Pleiades and Hyades types or in Trumpler's types 1 and 2.[27]

The Henry Draper stars in I. C. 2602, N. G. C. 3532, Messier 6, and Messier 7 (third column of Table III, VI), are arranged in order of spectrum and magnitude in Table III, VII. The large percentage of A stars in the Pleiades type of cluster is here seen in detail. It is to the frequency of the A stars in galactic clusters that we owe the attempts to determine spectral parallaxes for these systems on the basis of the material of the Henry

[27] *Ibid.* See footnote 7, Chapter II, on his later work.

Draper Catalogue.[28]   The wide dispersion in magnitude also shows why the attempts are not always happy.

TABLE III, VII —SPECTRA AND MAGNITUDES FOR SOUTHERN CLUSTERS

Cluster	Pg Mag	O	B0-2	B3-5	B8-9	A0-2	A3-5	F	G	K	M	All
I. C 2602	2		1									1
	3			1								1
	4			1								1
	5			3	1	1						5
	6			3	5	1						9
	7		1	2	3	6	2			1		15
	8				6	5		1		2		14
	9					7		1	1	3		12
	All	0	2	10	15	20	2	2	1	6	0	58
N G C 3532	5								(1)			(1)
	6											0
	7				3	2						5
	8			1	13	25	1	3	5	2		50
	9				7	43	1		2	2		55
	10					21				4		25
	All	0	0	1	23	91	2	3	7	8	0	135
Messier 6	6				1							1
	7		1	1	6	1		1				10
	8				7	11	1		1			20
	9				4	20		1	4	5		34
	10									10		10
	All	0	1	1	18	32	1	2	5	15	0	75
Messier 7	6				3	1				1		5
	7	1			14	18		1		3		37
	8			3	19	50	7	5	8	7	1	100
	9				2	15	2	2	2	6	1	30
	10					3			2			5
	All	1	0	3	38	87	9	8	12	17	2	177

b. Spectra in the Brighter Clusters.—The Pleiades,[29] the Hyades,[30] and Coma[31] fall under the head of "very loose and irregular clusters" (division c in the classification of galactic clusters in Chapter II).   References to discussions of their spectra and colors will be found in the special bibliographies

[28] Doig, J. B. A. A., **35**, 201, 1925; Raab, Lund Medd., Ser. 2, 28, 1922.
[29] See special bibliography in Appendix D and, for a detailed summary, Hertzsprung, M. N. R. A. S., **89**, 660, 1929
[30] See Appendix D.
[31] *Ibid.*

devoted to them. Figures III, 5 and III, 6 show, for the
Pleiades, the spectrum-magnitude and color-magnitude relations, the former compiled from the Henry Draper Catalogue
and Harvard Bulletin 764, and the latter taken from Hertzsprung's recent discussion.[32] The spectral data on h and

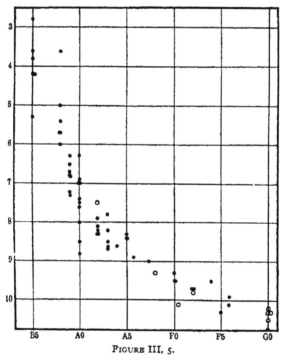

FIGURE III, 5.

Relation of spectrum to luminosity in the Pleiades. Ordinates,
photographic magnitudes, abscissae, spectral classes.

χ Persei, Messier 34, Messier 11, and two southern clusters are
summarized below.

1. The Double Cluster in Perseus.—The table of spectra
and magnitudes for the (combined) Perseus clusters has been

[32] M. N. R. A. S., 89, 660, 1929.

compiled from the data given by Trumpler,[33] but also includes some stars not given by him, which appear to belong to the systems, on the basis of proper motion, radial velocity,[34] or spectrum,[35] or more than one of these factors. The data are plotted in Figure III, 7, which shows also a number of the fainter stars that are plotted, but not individually listed, by Trumpler. Two matters of exceptional interest in this cluster are the fact that a large number of the brighter stars show the c-character and that the numerous bright-line stars are by no means the brightest members of the cluster.

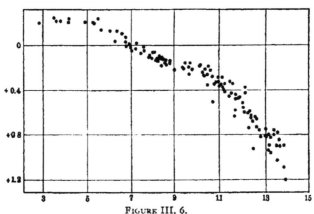

FIGURE III, 6.

Color-magnitude curve for the Pleiades. Color index (ordinate) is plotted against apparent photographic magnitude.

2. *Messier* 34.—The magnitude-spectrum relation for Messier 34 is illustrated in Figure III, 8, which is taken from Trumpler's discussion of the classification of open clusters.[36]

3. *Messier* 11.—In supplementing my earlier work on the colors in Messier 11, I found from plates made with the 100-

[33] P. A. S. P., **38**, 350, 1926.

[34] van Maanen, Dissertation, Utrecht, 1911; Pop. Astr., **25**, 108, 1917; Mt. W. Contr. 205, 1920; Adams and van Maanen, A. J., **27**, 137, 1913.

[35] Hertzsprung, B. A. N., **1**, 151, 1922; **1**, 218, 1923.

[36] Trumpler, P. A. S. P., **37**, 310, 1925.

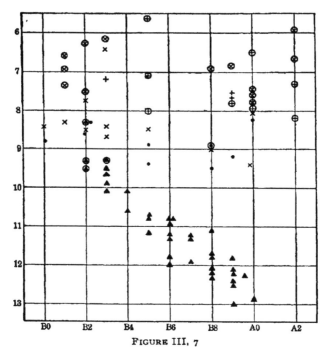

FIGURE III, 7

The magnitude-spectrum relation for stars in h and χ Persei.
Sources are as follows: +, Hertzsprung; ×, van Maanen; ▲.
Trumpler; ●, Payne. A circle enclosing the symbol indicates
a c-star.

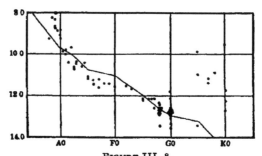

FIGURE III, 8.

Magnitude-spectrum curve for Messier 34.   Ordinates are apparent visual magnitudes, abscissae
are spectral classes.   The broken line represents
the dwarf branch. (After Trumpler.)

TABLE III, VIII —MAGNITUDES AND SPECTRA FOR THE DOUBLE CLUSTER IN PERSEUS

B D.	H.D	Spectrum	Magnitude	Note	B.D.	H.D	Spectrum	Magnitude	Note
+63°274	12301	cB5	5 50	1	+55°564	13970	B3	8 4	2
+57°494	12953	cA2	5 96	2	+56°500	14052	B5	8 5	2
+55°612	14818	B2	6 05	2	+56°498	14053	B2	8 5	2
+56°438	13267	cB3	6 19	2	+56°545	14250	B5	8 5	3
+56°471	13854	cB1	6 20	2	+56°485	13969	B2	8 6	3
+56°530	14143	Bo	6 42	2	+56°574		B4	8 7	4
+56°522	14134	Bo	6 42	2	+56°555	14357	B3	8 7	2
+57°519	13476	cAo	6 50	1	+56°577	14476	Boe	8 8	3
+56°568	14433	cA2	6 60	2	+56°527		B6	8 4*	2, 5
+55°588	14322	cB9	6 82	2	+56°478	13890	cB8	8 9	3
+56°593	14542	cB8	6 90	2	+56°479	13900	B5	8 9	3
+63°315	14010	cB5	6 93	1	+55°547	13561	B8	9 0	2
+56°470	13841	c B1	6 99	2	+56°565	14422	Bpe	9 2	3
+57°582	15497	B3	7 03	1	+56°535	14162	B	9 2	3
+57°568	14956	cB1	7 10	2	+56°588	14520	B9	9 2	3
+57°576	15316	cA2	7 36	1	+56°575		cB2	9 3	4
+56°621	14899	acAo	7 42	2	+56°578		cB2	9 3	4
+56°591	14535	c Ao	7 46	2	+55°596	14453	Ao	9 4	3
+54°539	14827	B9	7 49	1	+56°550	14321	B5	9 4	3
+56°475	13866	cB2	7 5	2	+56°445	13370	Ao	9 4	2
+57°522	13633	B9	7 63	1	+56°576		cB2	9 5	4
+59°535	16778	cB9	7 69	1	+56°507	14092	B8	9 5	3
+55°554	13745	B2	7 77	2	+56°571		B3	9 5	4
+57°634	17145	cB	7 8	1	+56°563		B3e	9 6	4
+57°526	13744	cAo	7 8	2	+56°573		B3e	9 9	4
+63°310	13590	cB5	7 9	1	+56°572		B4	10 1	4
+57°594	15963	acAo	7 98	1	(van M 380)		B3e	10 1	4
+56°543	14210	Ao	8 0	2	(van M 374)		B4e	10 6	4
+58°397	13412	acA2	8 3	1	+56°580	.	B5	10 7	4
+56°567	14434	B2	8 3	3	(van M 385)		B6	10 8	4
+57°525	13716	B1	8 3	2	(van M 384)		B6	10 8	4
+56°570	14443	cB2	8 4	4	(van M 510)		B5	10 8	4
+56°469	13831	Bo	8 4	2	(van M 413)		B6	10 9	4

NOTES TO TABLE III, VIII

1. Hertzsprung (B A N. 35 and 37) suspects that these stars are members
2 Included on the basis of the radial velocity given by van Maanen in Mount Wilson Contribution 205 Two stars of Class G are excluded.
3. Nearby stars suspected of membership from their spectra and magnitudes
4 Members of χ Persei enumerated by Trumpler (P A S P , 38, 350, 1926). Twenty-five stars that he plots but does not tabulate are omitted from this table.
5 Magnitude from the B. D.

inch reflector that the brighter spectra are chiefly of Class A.[37]
This observation was later verified by Lindblad[38] on other

TABLE III, IX.—SPECTRAL TYPES IN AND NEAR MESSIER 11

Stratonoff	Spectrum	Photographic Magnitude	Stratonoff	Spectrum	Photographic Magnitude
545	B7	8 29	523	A2	12 22
687	F3	9 50	631	A3	12 26
694	K0	9 85	452	A0	12 29
437	B8 5	10 36	662	A0	12 29
172	B?	11.40	356	A0	12 32
237	B9	11 51	381	A0	12 33
336	A0.5	11 55	158	B9	12 36
227	A1	11 56	493	B9	12 36
395	B8	11 58	143	B9	12 40
501	B9	11 60	462	A	12 40
516	B9 5	11 62	647	A1	12 47
581	B7	11 79	159	A0	12 47
331	A0	11 80	571	A0	12 47
345	B9	11 89	664	B8	12 47
577	A5	11 89	415	A2	12 54
235	A2	11 94	385	B5	12 61
663	A5	11 98	568	A0	12 62
686	A0	11 99	273	A0	12 66
232	A0	12 00	294	B9	12 66
407	A1	12 10	563	A4.	12 66
211	A1	12 14	193	A3	12 73
245	B9	12 14	188	A4	12 80
456	A0	12 14	267	F0	12 80
566	A4	12 14	698	K	12 83
544	A5	12 18	221	B6	12 84
375	A1	12 21	710	A2	12 98
705	A3	12 21	620	A2	13 05
197	A3	12 22	423	K2	13 31
740	B9	12 22			

[37] The magnitudes determined for this cluster are apparently in error, probably by a constant amount, notwithstanding the consistency of the Mount Wilson photometric plates (Mt. W. Contr. 126, 1917). The color indices are systematically too great, as shown by my own spectrum plates (unpublished) and subsequently by the similar work of Lindblad and Trumpler. A correction of −0.4 to the photographic magnitudes is indicated by an unpublished Harvard plate, but still the colors and spectra are inconsistent. There is a possibility of differential light absorption within the cluster.

[38] Mt. W. Contr. 211, 1918.

Mount Wilson spectrograms. The spectral classes of 59 stars in and around the cluster are given by Trumpler,[39] and his data are shown in condensed form in Table III, IX. Successive columns contain the number from Stratonoff's catalogue,[40] Trumpler's spectral class, and my photographic magnitude.

4. N. G. C. 3532 and N. G. C. 3766.—The distribution of spectra in two southern galactic clusters has been derived by

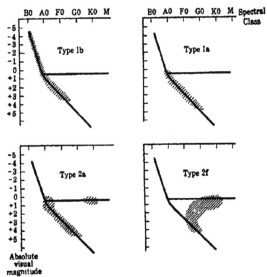

FIGURE III. 9
Trumpler's classification of galactic clusters.

Becker from his own photographs made at La Paz.[41] He finds percentages as follows:

	B0–B7	B8–A4	A5–A8	K0	K4	Number of Stars
N. G. C 3532	3 8	86.3	1.5	8.4	0	131
N. G. C. 3766	44 2	39.5	4.7	0	11.6	43

For N. G. C. 3532, Mrs. Fleming recorded 93 per cent of the 204 stars as Class A.[42]

[39] L. O. B., 12, 10, 1924.
[40] Tashkent Publ., 1, 1, 1899.
[41] A. N., 236, 327, 1929.
[42] See Table III, VI.

c. Trumpler's Investigations.—The kinds of spectral distribution among the stars of a galactic cluster, first recognized in the early Harvard work, and amplified by all subsequent studies, are defined in some detail in Trumpler's scheme of classification (Section 4 in Chapter II), and a number of galactic clusters are now assigned to the classes and their subdivisions. The essentials of the classification are shown by Figure III, 9, which is reproduced from Trumpler's paper.

The Lick studies, published and unpublished, have provided for the classification of 52 galactic clusters by means of their spectra, and the relative numbers in the various classes are as follows:

Type	Number
1b	24
1a	6
2a	20
2f	1
Others	1

It will be seen that the Pleiades type preponderates among the systems bright enough to classify. This may, however, be an effect of selection, as Type 1b contains far brighter stars than any of the others. The bearing of this selection on estimates of distance is considered later (Section 58, last footnote).

CHAPTER IV

VARIABLE STARS

THE most inviting and productive field in the study of globular clusters is the discovery and analysis of light variability. Fortunately, the field is rich. Almost a thousand variable stars are enumerated in the census contained in the present chapter. The first four sections refer to the many variables in globular clusters, the next one to their scanty appearance in galactic clusters, and the following sections deal with observational considerations of the nature of Cepheid variability.

16. A Summary of Known Variables.—Examining some of his earlier photographs of globular clusters, Dr. Common[1] noted the probable variability of some stars in Messier 5. Professor E. C. Pickering[2] in 1889 and Mr. David Packer[3] in 1890 also made some early observations of the variables in globular clusters, which were independently confirmed by Barnard a few years later.[4] But the whole development of this special branch of variable star astronomy is essentially due to Professor Bailey, whose extensive research on globular clusters, begun about 30 years ago, is the basis of much of our knowledge concerning cluster variable stars. Employing mainly the photographs made at Arequipa with various telescopes, Bailey has found the majority of cluster variables now known and has made by far the most important investigations of light curves and periods. Aside from Bailey's work, the discovery and

---

[1] M. N. R. A. S,. **50,** 517, 1890; **51,** 226, 1891.
[2] A. N., **123,** 207, 1889; H. C. 2, 1895.
[3] Sid. Mess., **9,** 381, 1890; **10,** 107, 1890; Engl. Mech., **51,** 378, 1890.
[4] A. N., **147, 243,** 1898.

study of the variables has been almost exclusively the work of Miss Woods at Harvard, Baade and Larink at Bergedorf, and the writer and his collaborators at Mount Wilson and Harvard (see references in Appendix C). Larink has made an extensive check of Bailey's periods for cluster-type variable stars in Messier 3, finding that, after a 20-year interval, 82 periods were unchanged and 29 probably had varied. My similar check on 54 of Bailey's variables in Messier 5 results in periods accurate to within a tenth of a second; in this cluster the periods are nearly all constant throughout an interval of 30 years.

The data at present available concerning the variable stars in globular clusters are summarized in Table IV, I. Clusters examined with care but without discovery of variable stars are also included in the table. Some stars suspected of variability are omitted in the absence of numerous or decisive observations. Three of these, for instance, are in the Hercules cluster,[5] where my measures on a few plates cannot be considered to furnish sufficient evidence of variability. In crowded regions and for close doubles, the photographic development (Eberhard) effect[6] may produce spurious variability, for it varies from plate to plate under ordinary working conditions. Aside from this and similar uncertainties, which lead to the inclusion and exclusion of suspected variables, there is an element of incompleteness in these tabulated results because of the difficulty of thoroughly examining the centers of clusters, and also because of the small number of plates sometimes involved in the surveys. In general, also, it may be said that scarcely a cluster has been examined with the accuracy and thoroughness necessary to detect ordinary eclipsing stars of short range or narrow minimum and to exhaust the possibility of Cepheids of small range.

[5] Shapley, Mt. W. Contr. 116, 79, 1915.
[6] Eberhard, Phys. Zeitschr., 13, 288, 1912; Pots. Pub. No. 84, 26, 1, 1926.

TABLE IV, I.—SUMMARY OF VARIABLES IN CLUSTERS

N. G. C.	λ	β	Class	Ellipticity	Variables	Period		Suspected	References
						$<1^d$	$>1^d$		
	°	°							
104	272	−44	III	8	7		3		1, 2
288	214	−88	X	9	2				3
362	268	−46	III	8	14				1
1851	211	−34	II	9	3				4, 5
1904	195	−28	V	9	5				1
3201	244	+10	X	9	61				6, 7
4147	227	+78	IX		5			8	8
4590	268	+37	X	9	28	27	1		9, 10
4833	271	−8	VIII	8	5				5
5024	307	+79	V	9	40				11, 12
5053	310	+77	XI	8	9	8			13
5139	277	+16	VIII	8	132	95	5		1, 14
5272	8	+77	VI	8	166	110	1	79	1, 15, 16, 17, 18 19
5286	280	+10	V	9 5	0	.			20
5466	8	+70	XII	9	14	12			12
5904	333	+45	V	9	84	69	3	8	1, 21
5986	305	+13	VII		1				1
6093	320	+18	II	10	4				1, 39
6121	319	+15	IX	9	33			5	22
6205	26	+40	V	9 5	7	1	2	3	1, 17, 23, 24
6229	41	+39	VII:		1				8
6266	320	+ 7	IV	8	26				1
6293	325	+ 8	IV	9	3				25
6333	333	+10	VIII	9	1				3
6341	35	+34	IV	8	14				17, 26
6362	293	−17	X	8	17				27, 28
6397	304	−12	IX	9	2				1
6539	348	+ 6	X	9	1				29
6541	317	−12	III	9	1				30
6553	333	− 4	XI	9	0			2	25
6584	310	−18	VIII	9	0				20
6626	336	− 7	IV	9	9				1
6656	338	− 9	VII	8	21	9	2	4	1, 31, 32, 33
6712	353	− 6	IX·	.	1				8

TABLE IV, I.—(*Continued*)

N. G. C.	λ	β	Class	Ellipticity	Variables	Period		Suspected	References
						<1^d	>1^d		
	°	°							
6723	327	−18	VII	9 5	17	16	0		1, 34
6752	303	−26	VI		1				1
6779	30	+ 7	X	8	1			2	25, 35
6809	335	−24	XI	9	2				1
6864	348	−28	I	9	11			5	25, 36
6981	3	−34	IX		29	29		5	8, 25, 36
7006	32	−20	I		11	11	0	.	25, 37
7078	33	−29	IV	8	74	60	1		1, 21
7089	22	−37	II	9	11		1		1, 38
7099	356	−48	V	9	3		.		1
7492	23	−65	XII	9	9			5	3

[1] Bailey, H A , 38, 2, 1902
[2] Bailey, H. B 783, 1923.
[3] Mt W. Obs , unpublished.
[4] Bailey, H. B. 802, 1924.
[5] Miss Swope, unpublished.
[6] Miss Woods, H C. 216, 1919.
[7] Bailey, H. C. 234, 1922.
[8] Miss Davis, P. A. S. P., 29, 260, 1917.
[9] Shapley, Mt. W. Contr 175, 1920. .
[10] Shapley, P. A. S. P., 31, 226, 1919
[11] Baade, Hamburg Mitt , 5, No 16, 1922
[12] Baade, Hamburg Mitt , 6, No. 27, 1928
[13] Baade, Hamburg Mitt., 6, No. 29, 1928.
[14] Innes, U. C 59, 201, 1923.
[15] Shapley, Mt. W Contr. 91, 1914.
[16] Shapley, Mt. W Contr. 176, 1920.
[17] Guthnick and Prager, Sitz d. Preuss. Akad d. Wiss., 27, 508, 1925.
[18] Larink, Bergedorf Abhandlungen, 2, No. 6, 1922.
[19] Barnard, A. N., 172, 345, 1906.
[20] Bailey, H. B 801, 1924

[21] Bailey, H A., 78, 1917.
[22] Miss Leavitt, H. C. 90, 1904.
[23] Russ Astr Journ , 1, 16, 1924.
[24] Shapley, Mt W. Contr. 116, 1917.
[25] Ibid , Mt. W. Contr. 190, 1920.
[26] Miss Woods, H. B. 773, 1922.
[27] Miss Woods, H. C. 217, 1919.
[28] Miss Woods, unpublished.
[29] Hubble, letter.
[30] Miss Woods, H. B 764, 1922, see also A. N., 215, 391, 1922.
[31] H B 848, 1927
[32] Zö-Sé Annals, 10, 1918.
[33] Bailey, Pop. Astr., 28, 518, 1920.
[34] Bailey, H. C. 266, 1924.
[35] Miss Davis, P. A. S. P , 29, 210, 1917.
[36] Mt. W. Contr. 195, 1920.
[37] P. N. A. S , 7, 152, 1921.
[38] Chèvremont, Bul. Soc. Astr de France, 12, 16, 90, 1898.
[39] Bailey, H. B. 798, 1924.

17. The Frequency of Variable Stars.—Although only 45 of the clusters enumerated in Appendix A have been thoroughly examined for variables, we may profitable indicate some preliminary results of the search.

a. Nearly 900 variable stars have been discovered in globular clusters, as enumerated in Table IV, I. The great majority of those for which periods have been derived are cluster-type variables with periods less than a day. Nineteen have been found, however, with periods greater than a day; reference is made to some of them in Sections 19 and 53.

b. Ten clusters are known to contain more than 25 variable stars each and may be considered rich in variables. The "poor" clusters, with less than 5 known variables, are 18 in number (excluding, of course, those that have been inadequately examined).

c. Richness is slightly correlated with galactic latitude; the mean latitudes for the 10 rich and the 18 poor clusters are 32° and 21°, respectively. The mean distances of the rich and the poor clusters from the galactic plane are 7,320 and 5,550 parsecs. Both these results, of which the latter is probably the more significant, suggest a correlation between poverty in variables and proximity to the galactic plane.

d. Figure IV, 1 illustrates the relation between the number of variables and the class of cluster. Rich clusters are confined to the intermediate classes; poor clusters are distributed equally throughout all classes. It seems unlikely that observational selection is responsible for this result, although possibly some part of the effect may be attributed to the decreased discovery chance in a very condensed cluster.

18. General Properties of Variables in Clusters.—A few somewhat disconnected points appear worth recording before the special characteristics of cluster-type variables are discussed. Much work remains to be done in discovery, in deriving colors, periods, and light curves, and especially in checking the earlier determinations of period (most of which are based on scanty material) for the study of the highly significant changes in periods and light curves.

a. The great majority of cluster variables are, on the average, 1.5 to 2.0 magnitudes fainter than the brightest stars in the cluster. The difference is evaluated numerically in Chapter XI for several clusters in which the variables have been studied. For others, where they have been found but not analyzed in detail, an inspection of the photographic plates and prints, such as those in Harvard Annals, **38**, supports this generalization concerning the relative brightness of the variable stars and the most luminous objects in the clusters.

b. In Miss Leavitt's survey of the Magellanic Clouds[7] no variables were found fainter than photographic magnitude

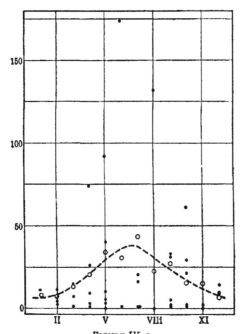

FIGURE IV, 1

Numbers of variable stars in clusters of various classes. Circles represent smoothed means.

17.5 (revised scale). The plates, made with the 24-inch Bruce telescope with exposures of from two to five hours, are sufficient to test this matter. A similarly definite fainter limit to the magnitude of Cepheids has been observed in globular clusters, particularly in ω Centauri, Messier 3, Messier 5, and Messier 13. No dwarf Cepheids are on record in clusters or the Galaxy.

[7] H. C. 173, 1912.

The negative results of the searches for faint variables and the form of the period-luminosity curve, which flattens conspicuously for periods less than a day, suggest that dwarf Cepheids do not occur.

c. There is as yet no convincing evidence of eclipsing binaries in any cluster.[8]   This is not very surprising, since in the Galaxy the great majority of eclipsing stars that show appreciable range are of Class A, with an average absolute magnitude of about $+1$.   They would therefore naturally be too faint for any studies so far made, except perhaps those of Messier 13 (where I made a special search on a few plates for faint variables), Messier 3, Messier 22, and ω Centauri.

d. The absolute magnitude of the variable stars, the period-luminosity curve, and the relation of the ordinary Cepheid to the cluster-type variable will all be further considered in Chapter X.   It may be remarked in passing that there is no evidence that the Cepheid variables in clusters, whether of short or long period, are different in their various characteristics from those in the Galaxy at large.   The color changes in cluster variables are found to be of the usual sort.   The light curves are of the uniform pattern.[9]   The long-period variables in 47 Tucanae are also normal[10]—their periods, ranges, probably their light curves are exactly duplicated by well-known galactic variables of the long-period class.

It is only in the relative numbers of different types and sub-types of variables that we find peculiarities in the globular clusters when contrasted with the galactic system, galactic star clouds, or the Clouds of Magellan.   These differences of content and frequency have not been explained satisfactorily. There are some indications that the Cepheids in the galactic system have properties that depend on environment, which are possibly the marks of differences in age and history.   It is to be noted, however, that very serious factors of incompleteness

[8] But see Appendix C, Ref. 202.
[9] See Figure IV, 2.
[10] Shapley, H. B 783, 1923.

and selection affect the comparison of clusters with the Galaxy, and also that there appear to be just as great differences in variable star content between clusters as there is, say, between ω Centauri or 47 Tucanae and the solar neighborhood.

19. Notes on Some Individual Variable Stars.—It is the less common variable with period greater than a day, rather than the cluster-type variable, that is bright enough to be studied in detail. The present section summarizes observation on such variables in five clusters.

a. Messier 3; Variable Bailey No. 95.—Bailey[11] found more than 100 variables in Messier 3, of which No. 95 is one of the brightest. It has a photographic magnitude of 13.9,[12] brighter than the value[13] (13.92) for the sixth star in the cluster. Our final modulus[14] of 15.43 for Messier 3 gives the photographic absolute magnitude of No. 95 as −1.5. The star has a large positive color index. Bailey was not able to derive a period from his observations.

Sanford[15] has obtained evidence of spectral variability for the star. When at maximum, it showed an apparently late-type spectrum with H γ and H δ bright; a month later, the star seemed to be fainter and the bright lines did not appear. A radial velocity of −300 kilometers per second was derived from the emission spectrum; Slipher's value of the velocity of the whole cluster is −125 kilometers per second.

Possibly the star is an irregular variable of the α Orionis type. Although emission lines are uncommon in the spectra of non-periodic red variables, they are sometimes found—for instance, in the spectrum of T Microscopii. It seems probable from the brightness of the variable that it is actually a member of the cluster and not a field star.

b. Bailey's Long-period Variable in Messier 22.—Among the

[11] H. A., **78**, 1, 1913.
[12] Shapley and Davis, Mt. W. Contr. 176, 1920.
[13] Shapley and Sawyer, H. B 869, 1929.
[14] Shapley and Davis, Mt. W. Contr 176, 1920.
[15] Pop. Astr , **27**, 99, 1919

cluster-type variables in Messier 22 Bailey[16] found a star with a period of 199^d.5. It has a large range, but the form of the light curve inclined the discoverer to regard it as a Cepheid rather than a typical long-period variable. The star is remarkable as being about a magnitude fainter than the other variables in the cluster—absolute magnitude about +1.0. This brightness places it with the long-period variables rather than with the brighter Cepheids (if, indeed, it is a member of the cluster), and even for that class it would seem to be rather faint. If not a cluster star, the variable must be in the background and very remote; the distance of Messier 22 itself is 6.8 kiloparsecs.[17]

   c. *Chèvremont's Variable in Messier 2.*—One of the brightest stars in Messier 2 was noted as variable by Chèvremont in 1897.[18] The star does not seem to have had further attention. The discoverer suspected a period of about 30 days, but the published observations suggest irregularity, which is borne out by estimates on 36 plates taken at Harvard and Mount Wilson. Throughout several years, a period of about 11 days is indicated. The range is rather more than a magnitude. With a photographic magnitude of about 12.5 at maximum, the star is two magnitudes brighter than the sixth star and, on the basis of the modulus[19] of 15.71, has an absolute magnitude of about −3. No data are available on color or spectrum. The star is within the recognized bounds of the cluster but may possibly not be a member.

   d. *Long-period Variables in 47 Tucanae.*—Of the seven known variables in 47 Tucanae, the three brightest are of long period:[20]

Period d	Maximum	Minimum
211.3	11 0	14.4
203	11 0	14 2
192	11 0	14.3

[16] *Ibid* , **28**, 90, 1920.
[17] Shapley and Sawyer, H. B. 869, 1929.
[18] Bul. Soc. Astr de France, **12**, 16, 90, 1898.
[19] H. B. 869, 1929.
[20] Shapley, H. B 783, 1923.

They are between one and two magnitudes brighter than the other four variables, for which no periods have been determined. From the adopted modulus for 47 Tucanae[21] (obtained without reference to the magnitudes of the variables) these long-period variables have an absolute photographic magnitude at maximum of −3.

e. *The Nova in Messier* 80.—Nova T Scorpii was discovered nearly in the center of Messier 80 (N. G. C. 6093) in 1860 by

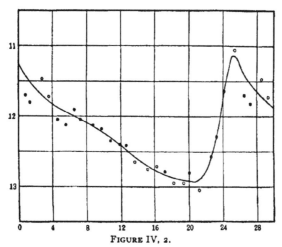

FIGURE IV, 2.

Light curve of variable No. 42, in Messier 5 Points with less than six observations are indicated by circles. Coordinates are photographic magnitudes and days.

Auwers (see Appendix C). If it is an actual member of the cluster, its absolute magnitude at maximum, when the apparent visual magnitude was 7.0, was −9.2, three magnitudes brighter than the average galactic nova.

f. *Messier* 5; *Variable* 42.—Illustrating the fact that the classical Cepheids in globular clusters are comparable with those of the Galaxy, a light curve of one of the variables in Messier 5 is given in Figure IV, 2.

[21] H. B. 869, 1929.

20. Variable Stars in Galactic Clusters.—Notwithstanding a careful search, no variable stars of recognized type are yet available for the estimation of the distances of galactic clusters. Variables have been found in the vicinity of Messier 11,[22] but probably they are members of the surrounding rich star field. The double cluster in Perseus was found by Lindemann[23] to contain variables, and others were suspected by Waterfield[24] in a special search of Harvard plates; but for none have periods been derived; all are probably irregular and may belong to the surrounding star field.

The only definite attribution of a regular variable to a galactic cluster is that by Doig;[25] he places T Serpentis, of maximum visual magnitude 9.0 and period $341^d.1$, in the cluster N. G. C. 6633. The star is probably about two magnitudes brighter absolutely than he assumed, and accordingly it is probably more distant than the cluster. Doig's spectral parallax of $0''.0025$ is nearly in accordance with the distance given in Appendix B.

21. The Relation of Magnitude to Period for Cluster-type Variables.—It is well known from the investigations of star clusters by Bailey and by the writer that the absolute median magnitudes for the short-period Cepheid variables in star clusters show very little dispersion. Furthermore, there is no dependence of median magnitude on length of period.

In the southern cluster ω Centauri, the subclasses a, b, and c of the cluster-type variables, with mean periods of $0^d.586$, $0^d.752$, and $0^d.395$, have median apparent photographic magnitudes of 13.55, 13.54, and 13.61, respectively, with an average deviation for a single star of one-tenth of a magnitude. The

[22] Barnard, A. J , **32**, 102, 1919; Pop. Astr , **27**, 485, 1919; Ritchie, P. A. S. P., **32**, 61, 1920; Walton, Pop. Astr., **35**, 25, 1927.
[23] Bul. St. Petersburg Acad Sci , Ser. 5, **2**, 55, 1895.
[24] Waterfield, W. F. H., unpublished.
[25] J. B. A A., **35**, 202, 1925.

mean of all (76 stars) is 13.57 ± 0.01. It appears that although the ranges and light curves differ conspicuously, as shown in Figure IV, 3, and the periods are in the approximate ratio of 3:4:2, the median magnitudes are, on the average, practically identical. Stars of different subclasses evidently are equal in total radiation. Whatever may cause the differences in period, amplitude, and light curve, it seems not to affect the average luminosity of the short-period Cepheids in this globular cluster, in sharp contrast to the conspicuous change of absolute magnitude with period for classical Cepheids.

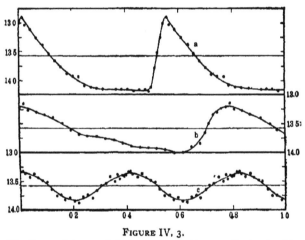

FIGURE IV, 3.

The three subclasses of cluster-type variables in ω Centauri   The coordinates are fractions of a day and photographic magnitudes. Observations for 5, 5, and 4 stars are used for subclasses a, b, and c, respectively.

Similarly, for Messier 3, Messier 5, and Messier 15, clusters for which there is material suitable for a quantitative test of the dependence of magnitude on period, the following means have been derived from Bailey's original observations. The data for ω Centauri are also included.

TABLE IV, II.—PERIOD AND MAGNITUDE FOR CLUSTER VARIABLES

Cluster	Number of Variables	Mean Period	Median Apparent Magnitude
		d	
Messier 3	1	0 32	15 53
	18	0 50	15.42
	18	0 54	15 53
	18	0 60	15 53
Messier 5	8	0 27	14 98
	10	0 48	14 98
	10	0 54	14 98
	10	0 63	14 96
Messier 15	29	0 36	15 66
	31	0 64	15.72
ω Centauri	c 34	0.39	13.61
	a 37	0 59	13.55
	b 19	0 75	13 54

The magnitudes of the variables in Messier 2, Messier 68, Messier 72, and other globular clusters yield results similar to the foregoing. In each cluster, apparently, the total light variation of the cluster-type variables is confined to a narrow interval of brightness, and the deviations of the median photographic magnitudes from their mean value are well within the errors of observation.[26] The median value is apparently an astronomical constant and, for this type of variable, a fundamental property. Its constancy throughout a considerable range of period length is a factor of some significance in the interpretation of the nature of the variation and, indeed, provides a crucial test for theory.

22. A Test for Hypotheses of Cepheid Variability.—

In all hypotheses of the cause of Cepheid variation that have

[26] There may be one known exception, No. 49 in Messier 15, which is six-tenths of a magnitude brighter than the 60 other cluster variables in the system for which periods are determined. But the star may be a component of an unresolved double, and therefore erroneously measured too bright, or possibly it is a foreground variable of normal brightness.

attained even partial success, the fundamental gravitational
relation between the period and the mean density, $P \propto \rho^{-\frac{1}{2}}$,
is generally accepted as applicable.[27] Variables of subclass
c in ω Centauri should, therefore, as a group have nearly four
times the density of those of subclass b. The extreme periods
in the cluster are $0^d.30$ and $0^d.90$, and in Messier 3, Messier 5,
and Messier 15 there is also a wide range of periods, the extremes
being

$$0.32 \text{ and } 0.71 \text{ for Messier } 3$$
$$0.23 \text{ and } 0.85 \text{ for Messier } 5$$
$$0.30 \text{ and } 0.76 \text{ for Messier } 15$$

Since for the periods P_1 and P_2 we have

$$\frac{P_1}{P_2} = \sqrt{\frac{\rho_2}{\rho_1}}$$

and, introducing surface area A and mass μ,

$$\frac{P_1}{P_2} = \sqrt{\frac{\mu_2 A_1^{3/2}}{\mu_1 A_2^{3/2}}}$$

it is clear that either the masses or the surface areas must change
considerably throughout the range of periods found among
cluster variables.

It is decidedly contrary to our beliefs and our present evidence that Class A stars of the same absolute magnitude should
differ in mass in the ratio of more than five to one. It is much
more reasonable to assume that the fixity of median magnitudes
and the general similarity of spectrum for cluster variables of
all periods mean essential identity of mass.

We are left, therefore, with the alternative of variety in the
sizes of cluster variables. But if we hesitate to accept large
and small cluster variables, we must then admit the failure of
the general relation connecting mean density and period.

[27] Shapley, Mt. W. Contr 92, 1914; 154, 4, 1917; Eddington, M N R. A. S..
79, 1, 1918; The Internal Constitution of the Stars, 192, 1926; Seares, Mt. W.
Contr. 226, 40, 1921; Jeans, M. N. R. A. S., 85, 808, 1925; Russell, Dugan, and
Stewart, Astronomy, 2, 766, 1927.

a. Assumption of Diversity in Diameters.—Let us consider the first possibility—diversity in size for stars of similar mass. Since for uniformity of mass,

$$P \propto A^{3/4}$$

the surface areas for cluster variables must differ throughout a range of at least four to one. To compensate for the increase of radiant surface with period, there must be a corresponding and exactly balancing decrease of surface brightness with period, amounting to 2.5 log 4 = 1.5 magnitudes in extreme cases and to 0.85 magnitudes in the case of the two distinct groups of cluster variables in Messier 15, for which the mean periods are $0^d.36$ and $0^d.64$, respectively. The large differences in surface brightness should correspond to conspicuous differences in spectral class and color, the shorter periods requiring bluer stars. The mean spectrum should differ by more than a whole spectral class and the color index by half a magnitude in extreme cases.

The observations on spectra for galactic short-period Cepheids are not extensive, but they certainly do not support this suggestion that the surface brightness and, therefore, the size of cluster variables may differ in such measure and progressively with period. The spectra for the individual stars vary on the average from Ao to F2. The median spectra show little dispersion.

For galactic cluster-type variables for which a single random determination of the spectrum is given in the Henry Draper Catalogue, we find

Mean Period d	Mean Spectrum	Number
0 38	A5 9	9
0 49	A5 6	8
0 59	A6 5	8

If we are to account for the constant luminosity by exactly compensating changes in diameter and surface intensity, the difference in spectrum between the first and last groups should be five units (corresponding to a difference in surface brightness expressed in magnitudes of $2.5 \log \dfrac{A_1}{A_2} = \dfrac{10}{3} \log \dfrac{P_1}{P_2} = 0.6 \Big)$, which is, however, eight times as large as observed and cannot be admitted.

b. Assumption of Differences in Color.—The best evidence on colors is derived from the two groups of cluster-type variables in Messier 15, for which Bailey determined the periods and I measured the average colors for each group, and from a similar but smaller list of variables in Messier 5, for which the colors are taken from my photovisual and photographic light curves of individual variables.

For one group of 31 variables in Messier 15 the periods range between $0^d.57$ and $0^d.71$, and for the other group, 29 variables, between $0^d.29$ and $0^d.44$. No periods between $0^d.44$ and $0^d.57$ were found, though in Messier 3 and some other clusters the periods of this length predominate.

The mean median magnitudes, as noted above, are the same for the two groups in Messier 15, within the errors of observation. This similarity holds also for the color indices, according to the measures of color made with the 60-inch reflector at Mount Wilson for 39 variables in the cluster.[28] A composite color curve was derived which showed that no important color differences exist between the two groups.[29] In summary, the results are

Mean Period	Number	Mean Magnitude	Mean C. I.	C I. near Maximum	C I near Minimum
Short (0^d 36)	22	16.00	+0 22	−0.14.	+0 36
Long (0^d 64)	17	16.04	+0.28	−0.02:	+0 35

The difference in mean period would require a change in mean color index of about three-tenths of a magnitude if the surface brightness were balanced against a hypothetical change in size with period.

[28] Shapley, Mt. W. Contr. 154, 14, 1917.

[29] Except, perhaps, near maximum where there is a suggestion that the variables with longer periods are redder (in keeping with results for Cepheids with periods longer than a day), but the observations are insufficient and uncertain at maximum where the long exposures are likely to decrease the observed range.

For Messier 5 the work on photovisual and photographic magnitudes to test for differences in the velocity of light[30] also gives results comparable to those above, showing the independence of color and period. The data are as follows:

Interval of Period d	Mean Period d	Number of Variables	Mean Color Index		
			Maximum	Median	Minimum
0 53 to 0 61	0 558	9	+0 24	+0 39	+0 60
0.45 to 0.50	0 475	8	+0 23	+0 43	+0.58

The systematic excess of two-tenths of a magnitude in all these color indices in Messier 5, which is more likely due to an error in the zero point of photovisual magnitudes than to a differential scattering of light in space, is of no consequence in the present comparison of colors for the two groups.

c. *Assumption of Nuclear Pulsations.*—From the foregoing results it appears that the hypothesis of differences in the colors and surface brightness, and therefore in the dimensions and mean density, for cluster variables with different periods cannot be maintained. The alternative of dispersion in mass has already been rejected. The only remaining course seems to be the admission of the failure for these variables of the simple relation between the period and the mean density.

It may be that the pulsation period should be associated not with the mean density of the whole star but with the density in the interior and that the various periods indicate merely stages in the development of a central nucleus which leaves the total mass, the *mean* density, and the size and brightness of the surface unaffected. This seems to me to be the only way out of a difficult situation.[31] And it, of course, requires for cluster-type Cepheids some modification of the pulsation hypothesis or other theory that may quite satisfactorily account for the classical Cepheids.

[30] *Ibid.*, H. Repr. 5, 1923.

[31] Eddington's more rigorous expression $P^2 = K/\rho f(\gamma)$, which introduces the ratio of the specific heats, would not help materially in reducing such a large anomaly except in so far as the specific heats may be involved in the suggested evolution of the nucleus.

One might attempt to connect, speculatively, these changes in period and nuclear readjustment with the hypothetical transition, in an evolutionary scheme, from the giant branch to the main sequence of stars, with the appeal to relatively quick gravitational adjustment after a particular material source of radiation has been exhausted.

In any case, the variables of the cluster type take on an added importance. They are linked with the problems of distances of clusters, Cepheid variation, and the interior structure and evolution of stars at the critical point between giants and dwarfs. It will be important in future studies to examine individual variables for sudden or gradual changes of period and of amplitude and to take the study of color into red and violet light. My preliminary "violet" magnitudes indicate for cluster variables smaller amplitudes than are shown by photographic light.[32]

23. Concerning Vestigial or Incipient Variation.—The most important irregularity in the general luminosity curve of

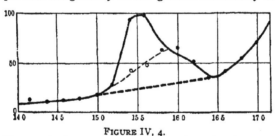

FIGURE IV, 4.

Photographic luminosity curve of stars of all colors in Messier 3 Ordinates and abscissae are numbers of stars and magnitudes. Exclusion of variable stars gives the broken line with circles

globular clusters lies at about the median magnitude of the cluster-type variables. It has been found for some clusters that the excess of stars at that point over an otherwise smooth luminosity curve is composed of blue stars. In Messier 3, it is found that the surplus stars are mainly the cluster variables

[32] Mt W. Contr 154, 16, 1917.

themselves, supplemented by other stars that resemble the variables in all characteristics except variability. The question naturally arises whether or not the light of the other stars that contribute to the excess is really constant. It may be that these objects are variables of small range, marking the beginning or ending of typical cluster variation.

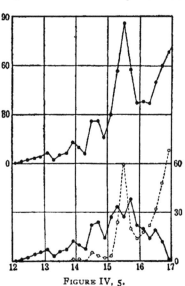

In a photometric catalogue of the brighter stars in Messier 3, which I published at Mount Wilson in collaboration with Miss Davis,[33] there are about one hundred typical cluster variables in the area studied, which does not include the dense and uncertain center. These hundred variables are only partially responsible for the hump in the luminosity curve, as may be seen from Figure IV, 4, which shows the photographic luminosity curve for all stars in the system from the fourteenth to the seventeenth magnitude

FIGURE IV, 5.

Luminosity curves in Messier 3. Above: all colors; below dots and full line refer to stars with color index greater than +0 60, circles and broken line, color index less than +0 60 Coordinates are numbers of stars and photovisual magnitudes.

If we omit all known variables and plot the luminosity curve for the remainder of the blue-white and the yellow-red stars separately, from the twelfth to the seventeenth magnitude, as in the lower part of Figure IV, 5, it is seen that the yellow-red stars do not contribute to the special maximum. (Since the photovisual magnitudes were used in this plot, the upper part of the figure gives, for purposes of comparison, the general photovisual luminosity curve, omitting variables.) The non-

[33] *Ibid.* 176, 1920.

variable blue stars, however, show the remarkable concentration between magnitudes 15.2 and 15.6. A smooth curve would allow less than 15 stars to this interval; the actual number is 83.

The stars that form the excess can be selected from the catalogue on the basis of color and magnitude. In the future, special attention should be directed to searching for possible irregularities in these objects. We already have a preliminary test of their stability in measuring the magnitudes for the general catalogue of Messier 3, for the size of the residuals from the several plates may be due to the variability of the stars as well as to the errors of observation. In brief, we obtain for the presumably invariable stars in the magnitude interval 15.2 to 15.6 (which includes the median magnitude of the variables, 15 3) the following results:

Number of Stars	Range of Color Index	Mean Photographic Residual
33	0 00 to +0 20	±0 103
19	+0 20 to +0 40	±0 083
37	+0 40 to +0.60	±0.071

The mean residual for all magnitudes and colors (800 stars) is ±0.082, but as this includes many values for stars that were later rejected because of uncertainty, it should be somewhat diminished for a fair comparison with the tabulated results. The interval of brightness near median magnitude was the most satisfactory of all for photometric measurement, being neither too bright nor too faint, and within it the accidental errors of measurement should be distinctly less than for the rest of the magnitude range. But the first line of the tabulation above shows the average residual to be abnormally large for the stars with colors and magnitudes like the variables. Moreover, two or three overlooked variables are not responsible for this large average deviation, for one half of the residuals exceed ±0.10. It seems reasonable to conclude that many of these stars are slightly variable and that we here witness the beginning or end of this type of Cepheid variation. At the maximum of the magnitude-frequency curve of these 33 stars the brightness is about a tenth of a magnitude fainter than the median magnitude of the variables.

24. A Composite Light and Color Curve for Cluster-type Variables.

—From a combination of photometric work done at Harvard and at Mount Wilson, it is possible to construct

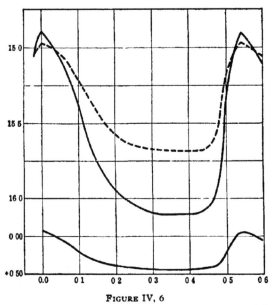

FIGURE IV, 6

Light and color curves for variables in Messier 3   Above, full line indicates photographic curve, broken line, photovisual Ordinates are magnitudes and color index, abscissae, fractions of a day

composite photographic, photovisual, and color index curves of 103 typical cluster variable stars in Messier 3. The result is essentially a highly accurate set of mean curves. Details of Bailey's work on the variables and my discussion and standardization of the magnitudes are published elsewhere.[34] The curves are shown graphically in Figure IV, 6, and numerically

[34] H. A., 78, 1, 1913; Mt. W. Contr. 154, 7, 1917; Mt. W. Comm. 70, 1920.

TABLE IV, III.—MEAN PHOTOGRAPHIC AND PHOTOVISUAL LIGHT CURVES OF
CLUSTER-TYPE VARIABLES IN MESSIER 3

Phase	Pg Mag	Color Index	Pv Mag.	Phase	Pg. Mag.	Color Index	Pv Mag.
0 000	14 90	−0 07	14 97	0 271	16 07	+0 42	15 65
0 011	14 93	−0 06	14 99	0 283	16 08	+0 42	15 66
0 023	14 96	−0 04	15 00	0 294	16 09	+0 43	15 66
0 034	15 00	−0.01	15 01	0 305	16 09	+0 43	15 66
0 045	15 04	+0 01	15 03	0 317	16 10	+0 43	15 67
0 057	15 11	+0 05	15 06	0 328	16 10	+0 43	15 67
0 068	15 17	+0 07	15 10	0 339	16 10	+0 43	15 67
0 079	15 23	+0 11	15 12	0 350	16 10	+0 43	15 67
0 090	15 32	+0 16	15 16	0 362	16 10	+0 43	15 67
0 102	15 39	+0 18	15 21	0 373	16 10	+0 43	15 67
0 113	15 48	+0 22	15 26	0 384	16 10	+0 43	15 67
0 124	15 58	+0 26	15 32	0 396	16 10	+0 43	15 67
0 136	15 67	+0 29	15 38	0 407	16 10	+0 43	15 67
0 147	15 73	+0 31	15 42	0 418	16 09	+0 43	15 66
0 158	15 79	+0 33	15 46	0 430	16 09	+0 43	15 66
0 170	15 84	+0 34	15 50	0 441	16 08	+0 42	15 66
0 181	15 87	+0 36	15 51	0 452	16 06	+0 42	15 64
0 192	15 91	+0 37	15 54	0 464	16 01	+0 40	15 61
0 204	15 94	+0 38	15 56	0 475	15 93	+0 38	15 55
0 215	15 98	+0 39	15 59	0 486	15 77	+0 33	15 44
0 226	16 01	+0 40	15 61	0 497	15 53	+0 24	15 29
0 237	16 03	+0 41	15 62	0 509	15 22	+0 11	15 11
0 249	16 04	+0 41	15 63	0 520	15 06	+0 02	15 04
0 260	16 06	+0 42	15 64	0 531	14 96	−0 04	15 00

in Table IV, III; in both places the adopted mean period is $0^d.54$.

For individual stars the light probably does not remain constant for three hours at minimum, as it appears to do for these mean curves. In other details, however, such as the forms, the amplitudes of 1.2 and 0.7 magnitudes, and the color variations, these results are essentially the same as those customarily derived for the average isolated cluster-type variable in the galactic system. In fact, the quantitative agreement of the very remote variables in Messier 3 with the local variables in such a phenomenon as color change is evidence of the accuracy of the photovisual and photographic magnitude scales relative to one another, and also of the absence of light scattering in space.

CHAPTER V

THE DISTRIBUTION OF STARS IN GLOBULAR CLUSTERS

A FULL knowledge of the distribution of stars in time and space is, of course, an unattainable dream. If we had, indeed, a clear picture of the ages, arrangement, and masses of the stars in even a single globular cluster, we should begin to know what is going on in this universe; but we are denied complete information. The distribution in time, or rather in stages of development, can only be inferred from fragmentary data on the magnitudes and colors of the brighter stars. The distribution in space is rather vaguely discernible through statistical conclusions based on laborious star counts. The distribution in masses, at least for giant and supergiant stars, is partially revealed in the general luminosity curves and the mass-luminosity relation.

Thus, for globular clusters we may eventually attain some idea of the distribution of mass and latent energy[1]; we shall probably have, at the most, an imperfect understanding of the distribution in space, and nothing but surmises about the distribution in time. A good preliminary surmise concerning stages of evolution is that all stars of a globular cluster originated at the same epoch and that their diversity in effective age is a consequence both of the dispersion in masses and of variety in dynamical experience. Other conjectures, however, are equally entertaining.

25. Are Cluster Stars Arranged Spirally?—In the early days of visual and photographic observation of star clusters it was natural that observers should look for a spiral arrangement of the stars and very frequently find it. A relationship between clusters and spiral nebulae was welcomed. The

[1] See Chapter XIV, Section 73.

coming of long-exposure photographs, which show in globular clusters tens of thousands of stars in nearly spherical arrangement, should have dispersed much of the interest in the arrangement of the few hundred or thousand brighter stars; but the subject has continued to attract attention.

Until recently, there has been a tendency to overemphasize the superficial arrangement of stars in both galactic and globular clusters. The laws of probability seem to be generally ignored in considering stellar distribution, for as soon as a slight departure from radial symmetry is noted among the brighter stars, we read of "lines of cleavage," "spiral paths," "lanes of nebulosity," and "channels of force." The significance that may be attached to chance groupings and chance vacancies, however, decreases remarkably with increasing exposure time. Nebulous obscurations that are reported to conceal the brighter stars are found, upon deeper penetration, to be ineffective for the more numerous faint stars, and therefore they cannot be real. Structural features other than flattening and central concentration may be present, but it is certainly inadvisable to conclude definitely, from knowledge of only a small percentage of the total number of the stars, that such structure exists. Probably in only one globular cluster (Messier 22) have stars as faint as the sun been photographed, and in only a few of those studied for stellar distribution have stars other than giants or supergiants been thoroughly examined

In order to determine whether the frequently described spiral structure in or near the center of globular clusters could be seen on large-scale photographs, I made a series of exposures on bright northern globular clusters some years ago with the Mount Wilson reflectors. The exposures varied in length. When only a few hundred stars were shown in a cluster, the spiral structure could almost invariably be traced; if the exposures were longer, the spiral arms became inconspicuous, or another set of arms, sometimes with different center and pitch, was found or imagined. The conclusion was reached that the phenomenon is wholly illusory. Spiral structure is the easiest form to visualize in centrally concentrated random groupings—especially when the number, pitch, thickness, origin, length, symmetry, and definiteness of the spiral arms are all arbitrary.

The *a priori* argument against the existence of spiral form in the images of globular clusters is, of course, simply that the clusters are three-dimensional. Cleanly traceable spiral arms would mean a most remarkable and unbelievable arrangement of stars in systems that are always nearly spherical. The discussion of the ellipticity of globular clusters in the next chapter will show how little they are flattened. Even the images of the much sparser galactic clusters are very rarely so elongated that we can assume them to be essentially two-dimensional.

Pointing out that vestiges of spiral structure or other persistent irregularities in globular clusters would indicate that the systems are not in a completely steady state, ten Bruggencate[2] has discussed evidences of spiral arrangement, and especially the data collected by Freundlich and Heiskanen.[3] In a discussion of our extensive Mount Wilson counts of stars in globular clusters, Heckmann and Siedentopf have recently concluded that there are no actual traces of spiral structure in the several globular clusters considered by them.[4]

26. On the Laws of Distribution.—The first detailed numerical consideration of the space arrangement of stars in globular clusters was made by Professor E. C. Pickering, who examined the distribution in ω Centauri, 47 Tucanae, and Messier 13[5] and proposed general empirical relations connecting surface density y with the distance from the center r in the forms

$$y = \int (1 - r^2)dz$$

and

$$y = \int (1 - r)^n dz$$

[2] Sternhaufen, pp. 63 ff., 1927.
[3] Zeits. f. Phys., **14**, 226, 1923.
[4] Gött. Veröff., Heft 6, 1929.
[5] H A., **26**, Chap XI, 1897.

Subsequently, much time has been devoted to studies of the laws of the distribution of stars in globular clusters. Various formulae relating the number of stars per unit volume N with distance from the center of the projected image r or with ρ, the distance from the cluster center, have been derived (or assumed) and applied to published counts of stars. We have, for instance, from von Zeipel,

$$N(\rho) = \frac{1}{\pi}\int_{\rho}^{R} \sqrt{(r^2 - \rho^2)}\frac{d}{dr}\left(\frac{1}{r}\frac{dn}{dr}\right)dr$$

where R is the radius of the cluster and n is the number of stars in the corresponding unit area of the projected image.

The problem of finding a law of space distribution from the law of apparent distribution in a globular star system has been solved by von Zeipel, who has been the leader in the attempt to deduce the structure of clusters from observation of stellar distribution. He was the first to utilize in this problem the principles of the theory of gases. Analogies with the kinetic theory have encouraged a number of theoretical and observational researches, but as yet no completely satisfactory representation of the observations has been found. Whether adiabatic cr isothermal distributions of gas most nearly simulate star distributions in globular clusters is not yet decided finally [6]

It seems unnecessary to treat in detail the history, methods, successes, and failures of these various investigations of distribution. No other phase of the study of globular clusters has been so frequently and thoroughly described. Special attention should be called, however, to the discussions by H. C. Plummer, ten Bruggencate, Strömgren, Eddington, Jeans, Parvulesco, and Martens.[7] A further important step has been made by von Zeipel and Lindgren, who proceed, from the assumption that the stars of different masses are distributed in equilibrium in relation to surrounding stars, to the determination, from the observed distribution, of the mean masses for

[6] See reference to Martens, Appendix C.
[7] See references in Appendix C.

various color classes and absolute luminosities. They have used the method with success in the study of the rich galactic cluster Messier 37, and Wallenquist has further discussed their analysis and applied the method to his own study of the magnitudes in Messier 36. Freundlich and Heiskanen have provisionally applied it to the study of the distribution of stars in globular clusters, but the observational material is yet too uncertain for satisfactory results.

That the discussions of the adiabatic or isothermal distributions of stars in globular clusters have been very unsatisfactory in practice is not surprising, since the data from which comparisons have been made are inherently faulty. The best chance for improvement and successful application of the theory lies in the few rich galactic clusters from which the foreground and background stars can be satisfactorily differentiated. For the globular clusters, however, the crowded centers and the attendant difficulties with Eberhard effect and background contrast vitiate the counts, except for the outer parts. Furthermore, the available counts deal with only the few hundred or few thousand supergiant stars; tens or hundreds of thousands of fainter stars, which must play a major role in stellar distribution, have not yet appreciably entered the investigations. Ten Bruggencate has recognized the importance of ellipticity in the distribution of stars in globular clusters, but otherwise all discussions of the problem have ignored this lack of radial symmetry.

The star counts that have been used for the study of globular clusters are almost exclusively those of Bailey at Harvard and of Pease and Shapley at Mount Wilson. Photographs on a larger scale are needed. Special attention should be given to the brighter and more open globular clusters (ω Centauri, N. G. C. 3201, N. G. C. 6397) and also to "giant-poor" anomalous systems and those clusters like N. G. C. 2477 that are possibly intermediate between the globular and the galactic types.

Tables and figures of the distribution of the stars in globular clusters are given for several of the brighter clusters by Pickering, Plummer, von Zeipel, and Heckmann and Siedentopf, to whose work reference is made in the bibliography in Appendix C. In conclusion, we must admit that the situation is not very hopeful. The frequency is (roughly) inversely proportional to the fourth power of the distance from the center. This holds only for giant stars in the typical globular clusters; the law of distribution of fainter stars is even less definitely known. We find them more widely dispersed than the giants, but we can

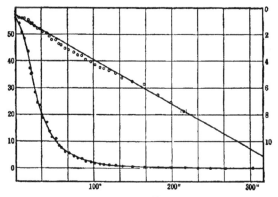

FIGURE V, 1.

Energy distribution in Messier 3. Abscissae, distance from center in seconds of arc. Ordinates, left and dots, energy per surface element in relative units, right and circles, in relative magnitudes (After Hertzsprung.)

say nothing of their distribution within the central sphere of ten parsecs diameter, where the crowding of brighter stars "burns out" the photographs.

An indication of the distribution of brightness in a globular cluster as a function of distance from the center is given in the curves of Figures V, 1 and V, 2, which show for Messier 3 and Messier 13 the distribution of light over the integrated images, as determined by Hertzsprung[8] and Hogg,[9] respectively. In both of these studies efforts were made to smooth out irregularities due to individual stars. The remarkable linear relation

[8] A. N., 207, 89, 1918.
[9] H. B. 870, 1929.

between distance and magnitude in Messier 3 is not found for Messier 13. The similarity in form of the curves for photo-visual and photographic magnitudes in Messier 13 is particularly noteworthy; the curves represent an inner region of the cluster that is largely excluded from my survey of magnitudes and colors.[10]

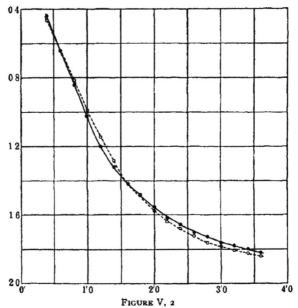

FIGURE V, 2

Energy distribution in Messier 13. Ordinates are magnitudes, with an arbitrary zero point, abscissae are minutes of arc, measured from the center. Circles and broken line represent photovisual magnitudes; dots and continuous line, photographic magnitudes

The distribution of light in the out-of-focus image of ω Centauri has been investigated by Schilt, who measured the photograph with a thermopile photometer.[11] He finds that the measured intensities show a much larger increase of stellar density toward the center than is revealed by Bailey's counts

[10] Mt. W. Contr. 116, 1915.
[11] A. J., **38**, 109, 1928; Pop. Astr., **36**, 296, 1928.

of the brighter stars.[12]   Nabokov's[13] study of Messier 13 gave results closely comparable with those of Hogg.

27. Luminosity Curves for Clusters.—Distribution in absolute brightness as well as in space can be satisfactorily studied as yet only for the giant and supergiant cluster stars. Although, as we have seen, the space distribution is not independent of the influence of dwarf stars, the fragmentary absolute luminosity curves have a meaning and a certainty that are essentially unimpaired by such forced neglect of the dwarfs. With the use of more powerful telescopes, moreover, especially on the borders of bright southern clusters in high galactic latitude, we shall soon be able to extend both the general luminosity curves and the luminosity curves for specific color classes to stars as faint as the sun or fainter.   In the near future we should therefore have for the globular clusters much more satisfactory data on the frequency of luminosities than we now have for stars of the galactic system.

The labor of determining magnitudes and colors on a satisfactory photometric basis is so considerable that luminosity curves will come slowly; for only three globular clusters are the color and magnitude surveys at all extensive.   It is necessary, therefore, to resort for the time being to general luminosity curves based on provisional magnitude scales, except for the giant stars in the three systems for which results are herewith presented.

a. Frequency Distribution of Giant Stars.—Observations for luminosity curves are listed in Chapter III above for the giant and supergiant stars of various colors in the globular clusters Messier 3, Messier 13, and Messier 22.   Corrections have not been made for superposed field stars; probably such corrections would not materially alter the forms of the luminosity curves.[14] Figure V, 3 shows graphically the combined results from the three clusters for six intervals of color class redder than f0.

[12] H. A , **26**, 213, 1897.
[13] Rus. Astr. Journ , **1**, 109, 1924.
[14] See Section 73 below for discussion of the field of Messier 22.

Stars redder than k5 are grouped together, since the data are insufficient for Class m alone; for classes b and a the observational limits of the three catalogues cut off the luminosity curves so quickly that the plots would not be significant. Probably Class f also is affected both by the confluence of giant and dwarf stars and by the magnitude limitations of the catalogues. The average dispersion is about half a magnitude, which is probably well in excess of the observational errors.

TABLE V, I.—COMPOSITE LUMINOSITY CURVE OF MESSIER 3, MESSIER 13, MESSIER 22

Absolute Photovisual Magnitude	Color Class						All Colors
	f0 to f5	f5 to g0	g0 to g5	g5 to k0	k0 to k5	>k5	
−4 0 to −3 5				1	1	1	3
−3 5 to −3 0	1			1		11	13
−3 0 to −2 5		2		5	13	18	38
−2 5 to −2 0	1	1	5	16	11	5	39
−2 0 to −1 5	1	7	34	32	6	2	82
−1 5 to −1 0	8	24	28	32	4	1	97
−1 0 to −0 5	18	85	71	20	1		195
−0 5 to 0 0	103	136	73	7	1		320
0 0 to +0 5	91	132	44	5	1		273
+0 5 to +1 0	38	31	8	1			78
+1 0 to +1 5	85	31	1				117
Totals	346	449	264	120	38	38	1,255

The coordinates of the mean luminosity curves are given in Table V, I. The tabulated quantities have been determined directly from the color-magnitude arrays (not from the original catalogues of magnitudes and color). Distance moduli used for reduction from apparent to absolute photovisual magnitudes are taken from Appendix A. The absolute photovisual magnitude limits of the catalogues are +0.4 for Messier 22, +0 5 for Messier 13, and +1.6 for Messier 3. The curves in Figure V, 3 and the numbers in Table V, I are therefore without much meaning fainter than $M_{pv} = +0.5$. The preliminary maximum at 0.0 for interval f0 to f5 is probably related to the humps in the general luminosity curves of these three clusters (Figure

V, 4), which are caused mainly by an abundance of blue stars and variables.

Until we have obtained data from other clusters, however, there is little point in fitting the usual exponential curves to the observations. Some of the asymmetries may be real in the magnitudes and disappear in the corresponding curves for total radiation or masses.

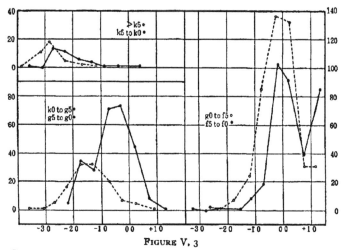

FIGURE V, 3

Luminosity curves for six intervals of color class, based on colors in Messier 3, Messier 13, and Messier 22. Coordinates are numbers of stars and absolute photovisual magnitudes.

b. The Preliminary Maximum.—General photographic luminosity curves based on provisional magnitude scales are given in Figure V, 4 for eight globular clusters. All the plotted material except that for N. G. C. 5053 is derived from my work on Mount Wilson plates. The methods of estimating the magnitudes and some of the numerical data are published elsewhere.[15] The curves suggest two general comments, which are followed below by a few special notes on the individual clusters.

[15] Mt W Contr. 155, 1917, 175, 1919.

1. There is a considerable variety in form of the general luminosity curve, an indication of the impropriety of using any method of parallax determinations that depends on the form of only the brightest portion of the general luminosity curve.

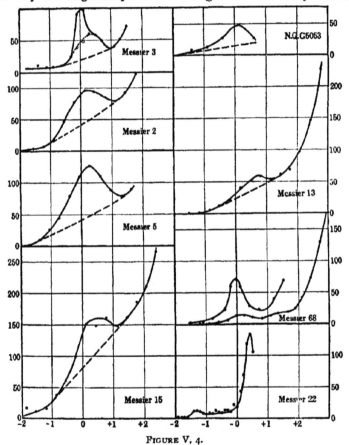

FIGURE V, 4.

Luminosity curves for eight globular clusters. Ordinates are numbers of stars, abscissae are photographic absolute magnitudes (approximate).

2. Without exception, all the globular clusters show preliminary maxima, which fall within half a magnitude of the median magnitude of cluster-type variables, as indicated by the two

heavy vertical lines in Figure V, 4. We have evidence in all of these globular clusters, though it is not shown graphically for N. G. C. 5053 and Messier 22, that the general luminosity curve rises steeply and high for stars fainter than the critical absolute luminosity near absolute magnitude zero. The bunching of stars at this particular brightness is probably of considerable significance in the economy of globular clusters. The resultant hump in the luminosity curves has possibilities in the measurement of distance.

    3. Details concerning Figure V, 4:

        (a) The preliminary maximum for Messier 3 is partially due to the 150 known variable stars; when the cluster-type variables are omitted from the diagram we have the milder and fainter hump shown by the broken line and small circles. The possible variability of these "hump" stars has been considered above in Section 23.

        (b) For Messier 2, Messier 5, and Messier 15 the variable stars have not been excluded, but they are not sufficiently numerous, even in Messier 5, to contribute much to the very conspicuous preliminary maxima.

        (c) The somewhat fragmentary luminosity curve for N. G. C. 5053 is derived from data published by Baade,[16] with variables excluded.

        (d) The preliminary maximum for Messier 13 is built up largely by stars of small or negative color index; the cluster has no definitely recognized cluster-type Cepheids.

        (e) For Messier 68 the two luminosity curves result from stars in different selected areas, measured on two plates. The curves are in fair agreement; a second preliminary maximum at $M = +1.4$ is suggested.

        (f) The fragmentary luminosity curve for Messier 22 is based on an unpublished catalogue, prepared at Harvard from my Mount Wilson photographs. The

[16] Hamb Mitt., 6, No. 29, 1928.

wide displacement of the maximum from absolute magnitude zero suggests that the correct distance may be 10 to 20 per cent larger than that given in Appendix A; but further work on magnitudes of the fainter stars is necessary before anomalies of the luminosity curve can be taken seriously.

CHAPTER VI

THE FORMS OF GLOBULAR CLUSTERS

THE perfect sphere—the form once commonly held to be an essential attribute of celestial bodies—is, according to observation and competent theory, a thorough illusion. There is exceedingly small chance of null rotation, and rotation is the end of sphericity. Oblateness, prolateness, and ellipsoidicity characterize planets, stars, nebulae, and galaxies. We now find that even the globular clusters belie their name. In a universe of moving particles the conception of the perfect figure vanishes with the growth of precision in measurement and attentiveness to detail.

Passing reference has been made in preceding pages to the ellipticity of globular clusters; Chapter VII will describe the irregularities of various sorts in the apparent forms of galactic clusters. The present chapter considers in some detail the deviations from circularity in the photographic images of globular clusters   The non-circular images imply, of course, deviations of the clusters themselves from sphericity.

28. Definitions and Difficulties.—The term "ellipticity" may be used in two senses, which are found to be practically equivalent:

1. On photographs, especially those of small scale, the integrated images of many globular clusters appear symmetrically elongated—the images are oval or elliptical.

2. In detailed star counts the density (number of stars) at various distances from the center is systematically greater along one axis of the projected image than along any other.

Both of these phenomena can be explained most naturally by assuming that the clusters are oblate; that is, the gradient of the space density of the stars differs in different directions from the center and is usually symmetrical about a polar axis and an equatorial plane. Probably the forms of the integrated photographic images are related as directly to star density as are the counts.

The ellipticity can be described in other terms, and possibly other explanations of the observed forms and star counts might be suggested; but in this chapter the foregoing interpretation will be adopted. Flattening may be a consequence of the rotation of the cluster or, in accordance with the theoretical work of Jeans,[1] a result of the encounter of the cluster with another stellar system; or both modes of deformation may be involved, especially in clusters which show irregular elongation.

It is difficult to determine the actual bounds of a globular cluster along various radii, or even the limits of the projected image, because of

1. The unknown and possibly peculiar density laws in different directions for a flattened stellar system.

2. The confusion with foreground stars.

3. The present lack, near the edges of individual clusters, of sufficiently long exposures, made with large reflectors.

The planes of symmetry in globular clusters may be likened to the galactic plane. Their discovery and measurement, however, is sometimes difficult. Three factors are involved in their detection—(1) the degree, (2) the orientation, and (3) the nature of the oblateness of figure which is revealed by the elliptical images.

1. The ratio of the shortest diameter to the longest may be near unity, so that only refined study, whatever the position of the observer in space (in or out of the cluster), would be able to detect the oblateness.

2. The cluster may be so oriented in space (polar axis nearly parallel to line of sight) that whatever the ratio of its major and minor axes the discovery of oblateness is impossible.

3. The absolutely brighter stars that enter our surveys may not show the oblate form, the property of concentration toward

[1] M. N. R. A. S., 74, 109, 1913; 76, 552, 1916; 82, 132, 1922.

a plane being confined to the fainter stars. This consideration
is very relevant. In our galactic system the naked-eye stars
of the later spectral classes show practically no galactic concentration. In globular clusters the brightest stars, which
through their arrangement have given the name "globular"
to the systems, are also mainly giants of the redder spectral
classes.

In some of the brightest and most thoroughly studied globular
clusters—Messier 13, Messier 22, ω Centauri—the ellipticity
is most pronounced. It is surprising, therefore, that the conspicuous elongation of globular clusters should have gone so
long undiscovered. Except for Bailey's star counts,[2] which
showed irregularities in the distribution of the brighter stars
for a few globular clusters, nothing very definite was known
of the important fact of deformation until the systematic star
counts on Mount Wilson photographs were analyzed.[3]

Bailey's counts, like the earlier visual and photographic
inspections, generally dealt with but one or two thousand stars
in each cluster. The underlying ellipticity, although not
confined wholly to the faint stars, usually becomes evident
only when large numbers of stars are considered. But the
numerous faint stars frequently make their presence known
en masse as well as in detailed counts, so that the diffuse integrated images of clusters can be used to study cluster forms
even if few or no individual stars are shown. The counts,
however, are frequently more searching. For example, in
Messier 13, described below, the Harvard plates do not show, on
inspection, the ellipticity revealed by the counts; they indicate,
if anything, a slight elongation in a direction 45° from the
major axis. The same is true of some other clusters. When
an integrated image is clearly elliptical, the numerical ellipticity
shown by star counts is very large. In a condensed cluster,
the Eberhard effect and other difficulties may interfere seriously
with the determination of the frequency of stars as a function

[2] H. A., **76**, No. 4, 1915.
[3] Pease and Shapley, Mt. W. Contr. 129, 1917.

of the distance from the center; but errors arising from such sources do not affect measures of the major axis of the cluster.

29. The Elongation of Messier 13.—Among the various globular clusters whose forms were investigated by the writer, partly in collaboration with Mr. Pease, Mrs. Shapley, and Miss Sawyer, the Hercules cluster, Messier 13, best illustrates the nature of the ellipticity throughout a wide range in magnitude.

The stars in Messier 13 were counted on nine plates, and the results arranged for analysis in a framework of 12 equal radial sectors and a series of concentric rings.[4] The plates are of various exposure times, ranging from one minute to five hours. The counts of the total number of stars in each of the 12 equal sectors show the amount of ellipticity and its change with magnitude. The counts for each of the zones between concentric circles show how the degree and direction of ellipticity varies with distance from the center of the cluster.

In Table VI, I the numbers of stars are given for six different plates for each sector separately, but with the zones not differentiated. For the two photographs of longest exposure Table VI, II again gives the number of stars in different sectors but divides the material into four zones for each plate. The central regions are too much burned out to be included in the star counts.

Three figures illustrate the evidence, derived from these star counts, for the ellipticity of Messier 13: Figure VI, 1 shows for

TABLE VI, I.—ELLIPTICITY FOR DIFFERENT EXPOSURES IN MESSIER 13

Duration of Exposure	Total Number of Stars	Numbers of Stars in Sectors											
		15°	45°	75°	105°	135°	165°	195°	225°	255°	285°	315°	345°
m 6	5,800	163	149	211	235	214	213	154	154	174	256	235	198
15	7,700	296	264	433	396	420	314	284	230	259	401	309	305
22	14,150	734	672	744	859	814	638	569	583	684	825	852	738
37 5	16,600	749	770	913	1,008	1,026	853	779	804	974	1,011	963	763
94	25,000	1,261	1,340	1,475	1,590	1,580	1,431	1,338	1,343	1,486	1,590	1,580	1,361
300	30,000	1,126	1,234	1,254	1,416	1,463	1,232	1,079	1,085	1,187	1,300	1,368	1,258

[4] *Ibid.*

TABLE VI, II.—ELLIPTICITY AND DISTANCE FROM CENTER IN MESSIER 13
(Results from Two Plates)

Distance from Center	Number of Stars in Sectors											
	15°	45°	75°	105°	135°	165°	195°	225°	255°	285°	315°	345°
2 to 4	623	668	750	762	764	728	712	683	758	778	718	670
4 to 6	358	361	423	464	476	386	330	352	402	438	479	394
6 to 8	168	207	212	236	226	202	188	194	214	248	249	198
8 to 10	112	104	90	128	114	115	108	114	112	126	134	99
3 to 5	560	640	624	684	788	642	586	597	629	658	712	662
5 to 7	362	374	410	471	431	360	302	292	358	396	424	394
7 to 9	204	220	220	261	244	230	191	196	200	246	232	202
9 to 11	141	136	116	118	130	129	131	124	130	157	134	100

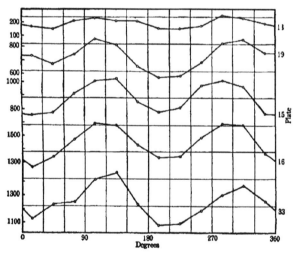

FIGURE VI, 1.

Axis of symmetry of Messier 13, shown by frequency curves for
five plates Coordinates are numbers of stars and position
angles.

five plates frequency curves which permit the determination of
the position angle of the major axis of the cluster; Figure VI, 2
gives the amount and position angle of the elongation at differ-

ent distances from the center; Figure VI, 3 shows the relation of the blue stars to the plane of symmetry.[5]

To summarize the results derivable from these tables and figures:

a. The cluster is conspicuously elongated, showing approximately 25 per cent more stars in the direction of the major axis than in the direction perpendicular thereto.

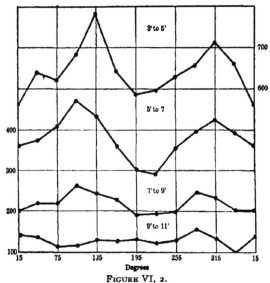

FIGURE VI, 2.

Ellipticity of Messier 13 for different intervals of distance from the center Ordinates are numbers of stars, and abscissae are position angles.

b. The elliptical distribution appears in all magnitudes from the thirteenth to the twentieth and fainter, though it is relatively inconspicuous for the stars brighter than the fifteenth magnitude. The ellipticity also shows throughout the cluster from center to edge.

c. A close analogy with phenomena of galactic concentration in our own stellar system is the decided preference shown by the

[5] Shapley, Mt. W. Comm. 45, 1917.

cluster stars with negative color indices for the sectors containing the major axis (Figure VI, 3).   The Cepheid variables in the cluster also lie near the major axis.

30. Ellipticity of Globular Clusters.—Over half a million stars have been counted in the course of the Mount Wilson and Harvard studies of the forms of globular clusters.   By means of star counts the position angles of the major axes have been found for 12 systems, and evidence of approximate circularity or asymmetry adduced for a few others.[6]   Direct estimates of

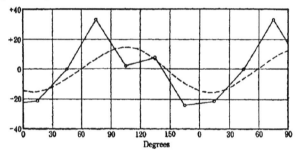

FIGURE VI, 3.

Relation of blue stars in Messier 13 to plane of symmetry.   Full line and circles refer to stars of negative color index; broken line shows the distribution of 10,000 stars between magnitudes 17 and 19   Coordinates are percentage deviations from the mean number of stars and angles of direction from the center.

the position angles for the integrated images of most of these clusters are found to agree closely with those determined from the laborious counts; hence, further surveys of the forms have been made with only the direct estimates of ellipticity on small-scale plates.   The work is described in the present section, and the chapter ends with a section dealing with a number of systems on which special studies have been made.

a. Counts and Estimates.—In the first systematic study of the forms of photographic images of the globular clusters, results were obtained for 41 systems—all that were bright enough for satisfactory examination on the Franklin-Adams

[6] Shapley, H., and Martha B. Shapley, Mt. W. Contr. 160, 1918.

star charts.[6] For 30 of the clusters the elongation seemed
sufficiently definite to justify a computation of the inclination
of the major axis to the galactic circle. For the brighter clusters, for which counts of individual stars had been made on
Mount Wilson plates, we compared the estimates of the
position angles of the major axes with those derived from
the counts.

TABLE VI, III.—COMPARISON OF COUNTS AND ESTIMATES

N G C	Messier	Position Angle of Major Axis		
		Counts	Estimates	Difference
		°	°	°
5024	53	160	170	−10
5139		105	120	−15
5272	3	Asym.	Asym.	
5904	5	55	50	+5
6121	4	115	Ind.	
6205	13	125	130	−5
6266	62	75	65	+10:
6273	19	15	15	0
6402	14	110	70	+40.
6626	28	50	45	+5
6656	22	25	25	0
7078	15	35	20	+15
7089	2	135	150	−15

The difference in position angles determined by the two
methods appeared satisfactorily small for these brighter clusters, but a later investigation,[7] made on Harvard plates, shows
that for the faintest clusters the estimates of the direction of the
axis are less certain than was believed from the study of the
Franklin-Adams charts.

b. *Orientation of Major Axes.*—The ellipticity and orientation given in columns 13 and 14 of Appendix A are based on the
newer Harvard investigation, except for starred values which
are taken from the earlier Mount Wilson counts. The ellip-

[7] Shapley and Sawyer, H. B. 852, 1927.

ticity is expressed as ten times the ratio of the minor to the major axis. The orientations, with respect to the galactic circle, are given only for those 39 clusters where the ellipticity is 8 or less, or if 9, only when the orientation could be certainly determined.[8] The orientation with respect to the galactic circle is reckoned from the "galactic east" (direction of increasing longitude) through the "galactic south"; negative angles consequently indicate reckoning from the east in the opposite direction.

c. *Inclination to Galactic Circle.*—The data bearing on the relation of the elongation to the plane of the Galaxy are collected for 37 globular clusters in Table VI, IV. (N. G. C. 1866 and N. G. C. 1978, in the Large Magellanic Cloud, are omitted from the table.) Successive columns are the number, class, galactic latitude, distance above or below the galactic plane in kiloparsecs, the degree of elongation, and the inclination of the major axis to the galactic circle.

Although the estimates here presented are probably better than any previously made, I feel that they are not of much significance except in the few cases where the ellipticity is conspicuous and the orientation angle is given without a colon. The average deviation of a position angle from the values determined earlier on Franklin-Adams charts is about 30°— an indication of the inherent difficulty of the estimates. Some of the clusters are definitely asymmetrical. For most of those not listed in Table VI, IV the deviations from circularity are small, or the clusters are too faint or too involved in a star field for useful determinations of form. Future star counts will probably find the axes not now revealed. Notwithstanding the difficulties and uncertainties, for all but 18 of the 93 clusters in Appendix A the degree of ellipticity is estimated.

In the future, attention should be paid not so much to small and difficult objects as to closer analyses of the distinctly asymmetrical systems and of the brighter clusters in which the degrees of ellipticity can be studied with respect to colors and

[8] That is, position angles without a colon in H. B. 852.

TABLE VI, IV.—ORIENTATION OF GLOBULAR CLUSTERS

N G. C	Class	Galactic Latitude	Distance from Galactic Plane	Degree of Elongation	Inclination to Galactic Circle
		°	Kpc.		°
104	III	−45	− 4 8	8	−55
362	III	−47	− 9 4	8	+65
1851	II	−34 5	− 8 1	9	−75
1904	V	−28	− 9 6	9	+ 5
2298	VI	−15	− 6 9	8	+39
2419	VII	+23	+11 9	9	−56
2808	I	−11	− 3 1	8	+84
4833	VIII	−8 5	− 2 2	8	−80
5024	V	+79	+17 9	9	−79
5053	XI	+78	+17 0	8	−61
5139	VIII	+15	+ 1 8	8	+30
5272	VI	+77 5	+11 9	8	+54
5897	XI	+29	+ 8 0	8	−44
5904	V	+46	+ 7 8	9	+16
6101	X	−16	− 5 7	8	+35
6121	IX	+15	+ 1 9	9	+72
6139	II	+ 6	+ 3 1	9	−64
6144	XI	+15	+ 4 8	8	−22·
6205	V	+40	+ 6 6	9 5	−63
6235	X	+12	+ 5 9	8	+89
6266	IV	+ 7	+ 2 3	8	+16
6273	VIII	+ 9	+ 2 6	6	−28
6341	IV	+35	+ 6 4	8	+16
6356	II	+ 9	+ 7 8·	9	−14
6362	X	−18	− 4 7	8	+78
6397	IX	−12 5	− 1 2	9	+73
6402	VIII	+14	+ 4 8	9	+76
6440	V	+ 2	+ 1 7:	8	+10·
6441	III	− 6 5	− 2 2	8	+40
6517	IV	+ 6	+ 5 2·	8	− 4
6626	IV	− 7	− 2 0	9	+18
6638	VI	− 7 5	− 3 9	8	−27
6652	VI·	−13	− 5 3	8	−11
6656	VII	− 9	− 1 1	8	+18
6779	X	+ 8	+ 2 8	8	+12.
7078	IV	−28	− 6 1	8	+11
7089	II	−36	− 8 2	9	−80

magnitudes. From detailed investigations, as von Zeipel has
shown, we may hope to get information on the masses of stars of

different colors and luminosities, using as criteria of mass the distribution with respect to the centers and to the hypothetical galactic planes within the clusters.[9]

31. The Relation of Elongation to the Galactic Circle.— The inclination of the axis of Messier 13 to the galactic circle is more than 60°; the inclination of Messier 22 is less than 20°. The former is in high latitude, the latter is the nearest of all globular clusters to the galactic plane. The question arises whether or not the galactic system has influenced the orientation of the globular clusters. In order that the equatorial plane of a cluster be parallel to the plane of the Galaxy, parallelism of the major axis with the galactic circle is a necessary though not a sufficient condition. We need to know the true form, or an equivalent, and get another component of the inclination, before we can fix the plane of the cluster in space; at present there is little prospect of measuring the true oblateness.[10]

The estimates of orientation given in Table VI, IV are admittedly uncertain, but the measures show definitely that large and small values are scattered throughout all distances from the galactic plane. There seems to be a slight tendency in the mean toward smaller values of the orientation for small values of $R \sin \beta$. Thus, we have:

Mean $R \sin \beta$	Number	Mean Inclination °
12 3	7	57
7 7	5	37
6 1	5	43
5 0	5	34
3 5	5	53
2.2	5	43
1 4	5	34

Grouped in order of inclination to the galactic plane, the same data give:

[9] Freundlich and Heiskanen, Zeit. f. Phys., **14**, 226, 1923.
[10] See discussion by ten Bruggencate, Sternhaufen, **72**, 1927.

Mean Inclination $^\circ$	Number	Mean $R \sin \beta$
82	5	7 5
75	5	4 1
62	5	9 6
46	5	6 6
28	5	3 8
17	5	3 9
10	7	5 5

The clusters in the direction of the galactic nucleus ($\lambda = 327°$, $\beta = 0°$) have lower average inclinations than elsewhere, according to the following means:

Mean Galactic Longitude $^\circ$	Number	Mean Inclination $^\circ$
39	6	39
203	5	52
277	6	52
308	5	64
321	5	36
331	5	30
343	5	27

The equatorial planes in the clusters with low inclinations of course may or may not be parallel to the galactic plane; those with high inclinations are certainly not parallel.

The class of cluster appears to be unrelated to the inclination of its axis to the galactic circle. Indeed, we may summarize this investigation of the orientation of the axes of globular clusters with the statement that there is no strong evidence that they are not inclined at random in space.

32. Seven Peculiar Clusters.—A few globular clusters that depart considerably from a spherical form have been investigated. They are of interest in connection with the possible dissolution of clusters, and especially in considerations of equilibrium and distribution laws.

a. Messier 62 (N. G. C. 6266).—The asymmetry of Messier 62 was particularly noted by Sir John Herschel[11] in 1847, and

[11] Cape Results, 23, 1847

later by Bailey.[12]   It is shown numerically in the following counts based on a Mount Wilson photograph, where numbers are given for 30° intervals of position angle, a central burned-out area one minute in diameter being excluded:

Position Angle	15°	45°	75°	105°	135°	165°	195°	225°	255°	285°	315°	345°
Number of Stars	67	56	67	45	35	42	49	56	68	79	72	72

When opposite sectors are combined, the position angle of the major axis appears to be about 75°; the estimated values of the angle[13] are 70° and 55°.

Messier 62 is the most irregular globular cluster. Does absorbing nebulosity cause its apparent asymmetry? Or has collision or encounter deformed it?

b. Messier 3 (N. G C. 5272) and Messier 5 (N. G. C. 5904).— It is difficult to account for such pronounced and deep-seated asymmetries in globular clusters as those of Messier 3, Messier 5, and Messier 62. For Messier 3 the distance from other stellar systems is now, and probably for hundreds of millions of years has been, extremely great; the galactic latitude is +77°.5. There is no evidence of occulting matter in the cluster or in front of it that might be responsible for a spurious appearance of irregularity. The measure of asymmetry is illustrated by the following data,[14] based on the counts of two Mount Wilson plates:

Position Angle	15°	45°	75°	105°	135°	165°	195°	225°	255°	285°	315°	345°
Number of Stars	244	234	246	220	198	242	222	282	292	321	303	263

Further work should be done on this cluster, as the counts on different plates are not wholly consistent.

For Messier 5 we have comparable data from a long-exposure plate showing 15,000 stars, made by Pease and counted by Miss Davis,[15] the star numbers referring to a region between 3′ and 15′ from the cluster's center.

[12] H. A., **76**, 74, 1915.
[13] Mt. W. Contr. 160, 7, 1918; H. B. 852, 25, 1927.
[14] Pease and Shapley, Mt. W. Contr 129, 1917.
[15] Shapley and Davis, P. A. S. P., **30**, 164, 1918.

Position Angle	15°	45°	75°	105°	135°	165°	195°	225°	255°	285°	315°	345°
Number of Stars	924	1016	972	861	693	783	870	978	1037	955	907	952

The asymmetry revealed by the counts for these three clusters produces a surprisingly minor effect on the general appearance of the clusters on photographs (see Plate I). The same result enters for the strongly elliptical clusters—they generally appear essentially round (as Messier 13, whose ellipticity is called 9.5 in Appendix A on the basis of its integrated image), though the counts show from 25 to 100 per cent more stars along one radius than along another.

c. Messier 19 (*N. G. C.* 6273).— Figure VI, 4 and the estimated ellipticities in Table VI, IV show that Messier 19 is the most elongated globular cluster so far studied. The circular coordinates of the diagram are position angles; the radial coordinates are relative numbers of stars in each 30° sector. The position angle of the major axis is 15°, and along this axis there are more than twice as many stars as along the minor axis. Photographs show no evidence of a double nucleus. Since the galactic latitude is $+9°$, and the major axis is inclined less than 30° to

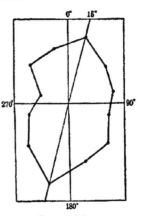

FIGURE VI, 4.
Diagram of star density in Messier 19.

the galactic circle, the equatorial plane of the cluster is probably nearly parallel to the galactic plane.

Considering this extreme example, we conclude that the flattening of clusters is systematically of a lower order than that of spirals and the galactic system.

d. Messier 22 (*N. G. C.* 6656).—The form of globular clusters near the galactic plane may have an important bearing on the relationship of the Galaxy and galactic clusters to the globular systems. Messier 22, in galactic latitude $-9°$ and one of the

nearest globular clusters, is significantly situated for a test of
deformation and the orientation of its central plane.   On a

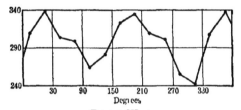

FIGURE VI, 5.

Distribution of stars in Messier 22.   Ordinates
are numbers of stars, abscissae are position angles

photograph which shows more than 70,000 stars, most of which
are undoubtedly members of the cluster though it is located in

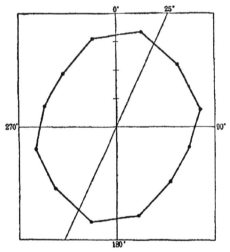

FIGURE VI, 6.

Star density for Messier 22.   Radial coordinates
show numbers of stars for 30° intervals of position
angle   The unit of scale indicated along the zero
axis is one hundred.

the direction of a rich galactic star cloud, a detailed count has
been made.[16]   The photograph shows stars from the twelfth

[16] Shapley and Duncan Pop Astr., **27,** 100, 1919.

to the twentieth magnitudes. The ellipticity is very pronounced, as illustrated by the curves in Figures VI, 5 and VI, 6, which refer only to the annulus with internal and external radii of $3'.6$ and $6'.4$. Near the center the counts would be unsafe because of crowding and occultation; too far from the center the cluster density falls off so that non-cluster stars would vitiate the count.

The number of stars in the direction of the major axis is nearly 30 per cent greater than the number along the minor

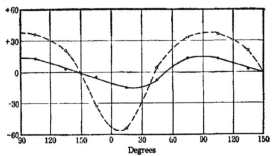

Degrees

FIGURE VI, 7.

The distribution of stars in ω Centauri. Ordinates are percentage excesses along the various radii indicated by the position angles (abscissae) The full line shows that the position angle of the major axis for all stars is about 90°. For the variables alone (broken line) the axis is about the same, but the relative excess is three times as great.

axis. The orientation of 18° and the high ellipticity, which suggests that we see the cluster edgewise, indicate that it may lie nearly parallel to the galactic plane.

  e. ω Centauri (*N. G. C.* 5139).—The 128 cluster-type variables in ω Centauri appear to lie preferentially along the equatorial plane of the system. Table VI, V and Figure VI, 7 illustrate this remarkable distribution.[17] The data are obtained from a Harvard plate, which shows the ellipticity for all distances from the center.

The numbers of variables in opposite sectors are combined in making the diagram. The relative amplitudes of the two

[17] Mt. W Comm. 45, 1917.

TABLE VI, V.—DISTRIBUTION OF 5,000 STARS IN ω CENTAURI
(Results from Two Plates)

Width of Zone	Number of Stars in Sectors												Mean
	15°	45°	75°	105°	135°	165°	195°	225°	255°	285°	315°	345°	
3 to 6	148	129	151	182	165	159	164	173	203	182	198	163	168
6 to 9	84	89	151	142	139	113	105	119	153	163	125	117	125
9 to 12	66	81	96	74	64	79	64	68	88	97	74	68	77
12 to 15	44	54	61	61	57	57	40	59	57	59	51	45	54
3 to 9	232	218	302	324	304	272	269	292	356	345	323	280	293
9 to 15	110	135	157	135	121	136	104	127	145	156	125	113	130
3 to 15	342	353	459	459	425	408	373	419	501	501	448	393	423
Variables	8	6	17	12	9	5	2	16	11	17	16	9	11

curves show that the variables are three times as condensed toward the supposed plane of symmetry as the stars in general, and the latter exhibit such strong ellipticity that ω Centauri is easily seen to be elongated on all photographs.

f. Messier 15 (*N. G. C.* 7078).—A series of long-exposure photographs on the edges of Messier 15 show that as far from the center as 15′, corresponding to a linear distance of 200 light years and quite outside the limits usually seen on the best photographs, the elongation of this cluster is in general agreement with the results obtained for the central region, both in direction and amount.[18]   The stars involved in the extension are extremely faint, and their frequency is, of course, relatively low.

[18] Pease and Shapley, Mt. W. Comm. 39, 5, 1917; Mt. W. Contr. 129, 11, 1917; Heckmann and Siedentopf, Gott. Veröff. Heft 6, 1929.

CHAPTER VII

THE STRUCTURE OF GALACTIC GROUPS

Turning from the smooth and symmetrical globular clusters to the typically irregular galactic groups, we note at once their strong contrasts. The galactic clusters are of various forms. The Pleiades suggested to the constellation makers Seven Virgins, or perhaps a Hen with Chickens; Praesepe was the Manger or the Beehive; the Hyades outlined the Face of the Bull. It does not take a lively imagination, however, to recognize the diversity and looseness which prevail in galactic clusters, while symmetry and central compactness characterize the globular systems.

The irregular boundaries make difficult the determination of the diameters and forms of galactic clusters; they are not easily untangled from the rich and fortuitously irregular star fields in which frequently they are embedded. Nevertheless, from star counts, studies of spectra, and measures of motion, it is possible to deduce the extent and form of a fair proportion of those now catalogued. In the following pages some of the more significant results are reported. A consideration of the distances of galactic clusters appears in Chapter XI, and of anomalies in their distribution in Chapter XIV; their numbers, classification, and distribution on the surface of the sky and data on their spectral composition have been taken up in Chapters II and III.

33. Elongation of Galactic Clusters.—A cluster of stars moving through a galactic field necessarily experiences transformation. It is affected both by encounters with individual stars and by the deforming forces of the whole galactic system. Jeans has shown that theoretically the form of a cluster at any time will depend upon its density and the length of time it has been bombarded by members of the star field which it penetrates.[1]   The galactic clusters (again in contrast with the globular clusters) are in low galactic regions where they are necessarily undergoing continuous perturbation. The observed irregularities of a cluster are therefore attributable not only to the chance arrangement of its relatively small population at a given time but also to various gravitational transformations and to the inevitable loss of individual stars. We naturally seek for evidence of some systematic effects of such changes in the orientation of the galactic groups.

a. Only for the comparatively rich and compact galactic groups is it possible to estimate the orientation, and even for them the degree of elongation cannot be determined with sufficient precision to be useful.   The loose systems—classes a, b, and c—are generally of too indefinite size and membership to show significant boundaries.   In the eighth column of the new Harvard catalogue of galactic clusters (Appendix B) are mean values, from two independent estimates, of the orientation, with respect to the galactic plane, of the major axes of 57 objects.   Twenty-four other clusters, of comparable richness and symmetry, were examined and the orientation found to be indeterminate (marked by "Ind" in the eighth column).   The two independent measures of the orientation agree surprisingly well, and there is no doubt that for most of these clusters the apparent elongation is correctly observed; but it is also possible that occasionally superposed non-cluster stars may be responsible for the appearance of elongation.

b. Of the 81 clusters suitable for examination 70 per cent show measurable orientations.   It is therefore appropriate to conclude that all are elongated, many so oriented as to avoid detection.   This is probably true, also, of both the loose groups and faint clusters for which no measurements were attempted.

c. The orientations of the major axes of the galactic clusters are best shown by Figure VII, 1, where lines representing the

[1] M. N. R. A. S., **82,** 132, 1922.

directions of the major axes are plotted on an equatorial system
of coordinates on which the galactic circle is also drawn.   The
mean value of the inclination is 44°.   The numbers of clusters
in successive 10° intervals of inclination, beginning with 0°
to 10°, are 6, 8, 4, 7, 10, 1, 11, 6, 4.

d. Taken as a whole, the 57 clusters appear to be oriented
at random, within the uncertainty of the determinations.   In
points of detail, however, we see that the average inclination
to the Galaxy differs from place to place, so that in some regions
the clusters appear to be mainly aligned with the galactic circle,

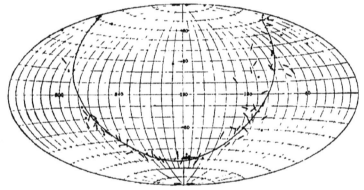

FIGURE VII, 1.

Orientation of major axes of galactic clusters, plotted on equatorial coordinates.
The black line represents the galactic circle.   Dots show positions of clusters
for which no elongation is found.

in others, oriented approximately at right angles.   The latter
condition is found, for example, in the clusters in Cassiopeia
and Perseus; in the opposite part of the sky, in the direction
of the galactic center, appears a decided tendency to parallelism
with the galactic circle.   The mean orientation for intervals
of galactic longitude (Table VII, I) further illustrates the irregularity but shows that we probably have little more than a random distribution of inclination.   The two high values at the top
of the table refer to the Cassiopeia-Perseus group of clusters,
and the last two entries to those in Sagittarius and Scorpio.

TABLE VII, I.—LONGITUDE AND INCLINATION

Mean Galactic Longitude	Number of Clusters	Mean Inclination
°		°
82 6	5	60
128 0	5	55
156 5	5	33
183 5	5	42
201 7	5	44
217 9	5	50
242 2	6	30
279 1	5	62
302 3	5	53
322 3	5	31
337 6	6	33

TABLE VII, II —INCLINATION AND DISTANCE FROM GALACTIC PLANE

Mean $R \sin \beta$	Number of Clusters	Mean Inclination	Mean Inclination	Number of Clusters	Mean $R \sin \beta$
		°	°		
677	5	45	80	7	69
425	5	38	70	5	203
265	5	47	66	4	167
196	5	34	59	7	279
143	5	51	47	5	215
100	5	52	40	5	131
62	5	47	29	6	281
30	5	39	18	5	175
17	6	53	12	4	222
6	6	47	4	4	108

e. The lack of correlation between orientation and distance from the galactic plane is illustrated in Table VII, II for the 52 clusters with definite values for inclination and $R \sin \beta$. The distances are expressed in parsecs.

The values of $R \sin \beta$ depend on preliminary estimates of the distances of galactic clusters. In means, however, they should be sufficiently dependable for the comparison with the inclinations.

34. The "Shoulder" Effect in Messier 67 and Elsewhere.—In spectral composition Messier 67 appears as a third type of galactic cluster, differing fundamentally from the Pleiades and Hyades models.[2] Its colors and magnitudes were the first to be systematically investigated in the Mount Wilson work on open clusters. Not only was the relation of magnitude to color found to differ from that in globular clusters, the Pleiades, and the Hyades, but also a peculiarity in the distribution of its stars was brought to light.[3]

TABLE VII, III.—AVERAGE STAR DENSITY, PHOTOGRAPHIC MAGNITUDE, AND COLOR INDEX IN MESSIER 67

Distance from Center	Number of Stars	Area in Square Minutes	Number Stars per Square Minute	Av Pv Magnitude	Average Color Index	Number of Stars	
						Background	Cluster
0 0– 2 5	35	19 6	1 78	12 49	+1 00	6	29
2 5– 4 5	64	44 0	1 45	12 85	+1 00	13	51
4 5– 6 5	49	69 1	0 71	12 52	+0 91	21	28
6 5– 8 5	30	94 3	0 32	13 03	+0 81	28	2
8 5–10 5	34	119 5	0 28	13 06	+0 82	36	(−2)
10 5–11 5	20	69 1	0 29	13 11	+0 80	21	(−1)
Total	232		.			125	107

A summary of the average star density, photovisual magnitude, and color index is given for 232 stars in Table VII, III. There is no decrease of star density or change of average magnitude or color, outside a circle of 6'.5 radius about the center of the cluster, but inside that circle the density, brightness, and redness of the stars increase towards the center. The number of cluster stars, contained in the last two columns, are obtained by assuming that the constant density outside the circle of radius 6'.5 refers to the background or foreground stars. The results indicate that less than half the measured stars belong to the cluster and that even within a radius of 6'.5 from the center, 27 per cent are not members of the system.

[2] Chapter II, Section 4; Trumpler, P. A. S. P., **37**, 316, 1925.
[3] Shapley, Mt. W. Contr. 117, 1916.

Although it seems conclusive, from this table, that the cluster is composed of about 100 stars scattered over an area of radius 6'.5, further study indicates that the evidence is misleading. The background density, deduced from the table as 0.3 stars per square minute, is nearly ten times as large as would be expected for the galactic latitude and photovisual magnitude concerned. This condition suggests that the cluster with radius 6'.5 is merely a well-marked nucleus of brighter and redder stars in a much larger system. To test the matter further, counts were made on Wolf-Palisa photographic charts of all stars within a degree of the center of the cluster. From this material, shown in Table VII, IV, it appears at once that

TABLE VII, IV.—COUNTS OF STARS NEAR MESSIER 67 ON WOLF-PALISA CHART

°	8^h40^m						8^h44^m						8^h48^m	Means
+11	6	10	8	12	8	7	7	13	12	7	7	15	14	9 7
	8	9	19	11	15	10	10	5	9	12	12	12	9	10 8
	12	11	12	10	10	10	10	6	12	7	8	13	10	10 1
	7	10	10	15	15	10	6	13	14	13	6	10	9	10 6
	9	12	19	16	13	17	11	8	9	14	16	18	12	12 6
	12	14	17	14	14	20	23	16	16	15	11	9	10	14 7
+12	8	11	11	6	14	38+	40+	42+	9	15	8	10	5	16 2
	11	11	15	18	13	29	80+	80+	36	13	6	6	8	25 1
	13	8	9	12	23	36	45+	50+	21	10	9	6	5	19 0
	9	12	7	11	9	24	24	19	17	15	14	11	6	13 7
	12	15	7	9	14	19	8	13	11	10	22	10	6	12 0
	7	14	6	13	14	12	9	17	7	6	10	10	15	10 8
+13	18	9	13	11	13	12	15	5	13	3	9	6	11	10 6
Means	10 2	11 2	11 8	12 2	13 5	18 8	22 2	22 1	14 3	10 8	10 6	9 7	9 2	

the cluster extends far beyond the limits of the plates used for the magnitude work. The real diameter may be as much as one degree. Beyond 15' or 20' from the center the ratio of background stars to cluster stars becomes large.

If the system of Messier 67 is roughly spherical in form, the space occupied by the nucleus is about a hundredth of the total volume of the cluster. The total membership is approximately 500 stars between photovisual magnitudes 10 and 15, but only the central concentration of 150 stars would attract attention in an ordinary survey of clusters. Without the nucleus the slightly concentrated residue might long have escaped discovery. Further research is needed to determine the number of cluster stars fainter than magnitude 15 and their distribution in space; present evidence indicates that such stars may be totally absent.

The "shoulder" effect found for Messier 67 appears to be rather common among galactic clusters and is a characteristic of high significance in their relation to the development of the Galaxy. In a preliminary study of several clusters, beginning in 1918, Trumpler found a shoulder effect for the Pleiades, Praesepe, and h Persei.[4] It has later appeared in studies of Messier 11, Messier 37, and other galactic clusters and is analogous, perhaps, to the wide scattering of faint stars shown in studies of the globular clusters. Messier 11 may be but a nucleus of the Scutum cloud.

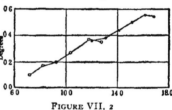

FIGURE VII, 2

Relation of radius to magnitude for h Persei Ordinates are distances from the center in degrees, abscissae, photographic magnitudes   (From Trumpler)

The dimensions of galactic clusters are much larger than they appear at first to be. From his studies of the distribution of faint stars in the vicinity of the Pleiades, Praesepe, and h Persei, Trumpler derived extreme diameters of 6°, 6°, and 1°, respectively, and observed, also, that the brighter stars are concentrated in the nuclei. A gradual change of radius with magnitude (outside the nucleus) is shown for h Persei in Figure VII, 2, based on data accumulated by van Maanen and Trumpler. For the Pleiades and Praesepe the "shoulder" effect appears to be more pronounced—there is a sharper contrast between nucleus and surrounding field.

35. Additional Remarks on Galactic Clusters.—Investigations now under way at the Lick and Harvard observatories and elsewhere will soon add so effectively to our knowledge of the structure and luminosity curves of galactic clusters that a complete summary of the numerous recent special investi-

[4] Allegheny Publ., 6, No. 4, 1922.

gations is inappropriate in this place. Attention may be called briefly, however, to a few studies which are indicative of the considerable activity in this field. References to the literature are given in Appendices C and D.

a. Messier 37.—Von Zeipel and Lindgren have made valuable contributions to the study of magnitudes and distributions in the rich galactic system Messier 37. From the space distribution for different magnitudes and colors they have determined, by the ingenious method devised by von Zeipel, the approximate

TABLE VII, V.—LUMINOSITY CURVES FOR VARIOUS COLOR CLASSES IN MESSIER 37

Limits of Magnitude	Color Classes					
	b	a	f	g	k	m
]11 0	0	2	0	0	0	0
11 0–11 5	0	4	1	1	0	0
11 5–12 0	0	12	0	5	0	0
12 0–12 5	7	21	0	10	0	0
12 5–13 0	8	18	0	1	0	0
13 0–13 5	20	16	3	0	1	0
13 5–14 0	5	37	2	0	0	0
14 0–14 5	3	29	7	1	0	0
14 5–15 0	1	19	16	3	1	0
[15 0	0	2	38	9	1	0

mean masses of the stars of various spectral classes. The method has been applied by Wallenquist to other galactic clusters, and Freundlich and Heiskanen have attempted to apply it to globular clusters, where, however, the available data were found to be as yet too incomplete.

The data of von Zeipel and Lindgren on luminosity curves for various color classes are given in Table VII, V. Tabulated quantities are numbers of stars. Incompleteness at the fainter magnitude limit is evident, but the results are definite in furnishing the complete luminosity curve for color classes b and g (giants) and in showing the significantly wide dispersion for stars of color class a. The double maximum of the luminosity curve for Class A stars has been noted as a possible means of estimating parallaxes   The method must be used with caution, however, because of possible confusion with chance irregularities when only small numbers of stars are considered.

b. *Foundation for Proper Motions.*—Küstner and Chevalier, among others, have made accurate modern catalogues of positions in several galactic clusters.   As a basis for future analyses of proper motions, this work is of high importance, since reliable measures of motions may soon be forthcoming for those galactic clusters that are relatively bright and near; for globular clusters the prospect is practically hopeless.   The catalogues of position also contain photographic magnitudes based on the international standards.

c. *Star Colors.*—Photographic and photovisual magnitudes of stars in Messier 35, Messier 36, and other northern clusters, as determined by Wallenquist, are appearing in a series of papers.   The results afford material for the discussion of bolometric magnitudes, luminosity curves, density laws, and probable mean masses, as well as star colors.

d. *The Shapes of Moving Clusters.*—The structure of well-known moving clusters and streams of stars, such as the groups in Taurus, Ursa Major, Scorpio, and Perseus, has been treated in some detail by B. Boss, Kapteyn, Eddington, Ludendorff, Hertzsprung, Bottlinger, and especially by Rasmuson, from the standpoint of measured motion.   The group motions are essentially parallel to the galactic plane.   Rasmuson shows that the Perseus and Ursa Major clusters are flattened at right angles to the direction of motion, the Scorpio group is flattened parallel to the galactic plane, and the Hyades cluster is nearly spherical.   These results on cluster forms are relevant to the investigation of the orientation of galactic clusters, reported in Section 33.   The approximate sphericity of the Hyades had been previously noted by L. Boss, and Turner first noted the flatness of the Ursa Major system.

e. *N. G. C. 7789.*—An investigation, made by Miss Mayberry and the writer, of the luminosity distribution of 5,000 stars

TABLE VII, VI.—STAR COUNTS FOR N. G. C. 7789

	Photographic Magnitude Interval					
	<16	16-17	17-18	18-19	19-20	Total
Cluster Stars	347	210	155	266	126	1,104
Field Stars	592	256	666	1,178	1,475	4,167
Ratio	0 59	0 82	0 23	0 23	0 09	

in the faint northern cluster N. G. C. 7789 has shown that in this cluster, as apparently in many others, the ratio of the number of cluster stars to the number in the field soon begins to decrease with decreasing brightness (Table VII, VI).[5] Apparently, most of the cluster stars have magnitudes between 15 and 19. It is well to observe, however, that this rapidly decreasing ratio does not mean the total absence of dwarf stars from the cluster, or even their relative scarcity. The increase in the number of field stars is probably but an indication of depth of field and the narrow space limits of the cluster. The additions to the field at any given apparent magnitude can be of varied luminosities; but additions to the cluster membership must be of specified absolute magnitudes.

[5] Mt. W. Contr. 190, 7, 1920.

CHAPTER VIII

ON THE VELOCITY OF LIGHT

THE astronomical determination of the absolute speed of light *in vacuo* has not been attempted in recent years because of its inaccuracy compared with measurements in the terrestrial laboratory where the work is subject to control and the reduction from atmospheric pressure to vacuum is simple and certain. The early procedure has, indeed, been reversed. The laboratory value of the velocity of light is adopted, and on this basis the eclipses of Jupiter's satellites, which were the first recognized indicators of the finite speed of light, serve now in the researches on masses and motions in the Jovian system.[1]

The classical laboratory method foreshadowed by Galileo and brought to the practical stage by Fizeau[2] and Cornu,[3] and the one attempted by Wheatstone,[4] which was developed by Arago[5] and Foucault,[6] were followed by the researches of Newcomb and Michelson. In 1905 the velocity of light was known with an accuracy of about 50 kilometers a second, or 0.000167 per cent. In 1924 the problem was taken up anew at Mount Wilson, under Michelson's skilful supervision.[7] The velocity is now placed at 2.99796×10^{10} centimeters a second. An accuracy to within 2 kilometers a second seems to be attainable.

In comparison with the laboratory accuracy, the astronomical measures, which depend on the diameter of the earth's orbit

---

[1] Sampson, H. A., **52,** Part 2, 1909.
[2] C. R., **29,** 90, 1849.
[3] Paris Obs. Annals, **13,** 1876.
[4] Phil. Trans. Roy. Soc., 1834, 583.
[5] Annuaire du Bureau des Longitudes, 1842, 287.
[6] C. R., **30,** 551, 1850; **55,** 501, 792, 1862.
[7] Mt. W. Contr. 329, 1927.

and the times of beginning and end of eclipses, have been of little value. There is, in fact, little prospect of determining accurately the *absolute* speed from astronomical measurements, but there is a preliminary test for the *relative* speed of light of different colors that can be made through the study of eclipsing stars; and by means of globular clusters an exceedingly accurate determination of the relative speed is possible, independently of the actual speed of light.

In these tests of the dependence of speed on wave length, which are of some importance in considering the nature of radiation, advantage is taken of large sidereal distances and, also, of certain properties of variable stars. We first consider briefly the preliminary indication from eclipsing stars and then turn to the more significant evidence provided by a remote globular cluster.

36. Stellar Eclipses in Light of Different Wave Lengths.—The Nordmann-Tikhoff effect has been studied in recent years by Y. and J. F. Cox,[8] by Russell, Fowler, and Borton,[9] and by others, none of whom has arrived at an unequivocal explanation of the observed difference in times of mid-minima for eclipsing stars when measured in light of different wave lengths. Originally, the lag was investigated in the hope and belief that it proved the absorption of light in space by interstellar gas; later, the fact was appreciated that any measurable difference in speed of light from nearby eclipsing stars would mean an impossibly large amount of dispersing material.

The measurement of the difference in the time of an eclipse in blue and in yellow or red light is an uncertain process, involving accurate light curves and freedom from systematic errors dependent on wave length. From a comparison of extensive photographic and visual observations made for six stars at the Harvard Observatory (photographic work by Miss Leavitt and visual work, except for W Ursae Majoris, by Wendell)

---

[8] Bul Obs Lyon, **9**, 113, 1927.
[9] Russell, Fowler, and Borton, Ap. J., **45**, 306, 1917.

Table VIII, I has been prepared from data computed by Russell and his collaborators. It shows in the fourth column the observed difference in time of mid-minimum, with its probable error. The estimates of distance in the third column are dependable enough for the computation of the differential velocities, which are given in the last column. Although the observed differences in time of mid-minimum are undoubtedly real for some of these stars, the resulting computations show such wide variance that the interpretation as differences in velocity cannot well be maintained. The differences in the fourth column for these six stars and for those discussed

TABLE VIII, I —ECLIPSING STARS AND THE VELOCITY OF LIGHT

Star	Period	Distance in Parsecs	Pg —Vis.	Velocity Difference in Meters per Second
	d		*d*	
S Cancri	0 485	420	+0 0085 ± 9	+5 1
W Delphini	4 806	890	+0 0026	+0 7
SW Cygni	4 573	735	+0 0042 ± 9	+1 4
U Sagittae	3 381	295	+0 0040 ± 5	+3 4
RW Tauri	2 769	675	−0 0025	−0 9
W Ursae Majoris	0 334	50	−0 0004 ± 7	−2 0

at Brussells[8] should probably be attributed to one or more of the following causes:

1. Accidental observational errors. (Frequently visual and photographic observations do not refer to the same minimum, and the difference is merely inferred from mean light curves.)

2. Uneven distribution of color over the surface of the eclipsing pair, for which there is good evidence for a few stars.

3. The systematic errors dependent on color, such as the humidity effect on photographic plates, which may make the beginning of an exposure much less effective photographically than the end and accordingly displace the mid-time of effective exposure from that recorded.

[8] Bul. Obs. Lyon, **9,** 113, 1927.

The Nordmann-Tikhoff effect certainly merits further careful study; meanwhile, the observations can be taken to indicate that the difference in the velocity of blue and yellow light must be exceedingly small. Table VIII, I shows, in fact, that the two velocities are the same to one part in a hundred million. In the following section we shall present much more conclusive evidence for equality in the speed of light, using quantitative material which comes incidentally from the study of globular clusters.

37. Messier 5 and the Relative Speed of Blue and Yellow Light.—For the accurate measurement of the relative speed of blue and yellow light we need (1) a source that emits light of various colors at controlled or predictable intervals, (2) a measured base line of great extent, (3) a device for receiving the signals sent over this base line in the form of light of different wave lengths. The requirements can be met by using the cluster-type Cepheid variables in distant globular clusters as the source of the light signals and receiving the signals simultaneously on blue- and yellow-sensitive photographic plates through appropriate light filters. My studies of the variable stars in Messier 5, resulting in the determination of photographic and photovisual light curves for a large number of individual stars,[10] have provided the material for the test described in the following pages.

The steep rise to maximum brightness of a typical cluster-type Cepheid variable makes the time of median magnitude on the ascending branch more accurately determinable than the times of maximum or minimum. Visual work by Wendell at Harvard on the star RR Lyrae and by the writer at Princeton on SW Andromedae was concentrated on the study of the rising branch of the light curve; it was found that the time of median magnitude can be determined to within 2 or 3 minutes. The maxima are usually sharp, however, and can also be timed with some success.

[10] H. B. 763, 1922; H. Repr. 5, 1923, P. N. A. S., 9, 386, 1923

The astonishing accuracy of the mean period of the cluster-type Cepheid led Barnard to propose the possibility of using the period of a star of this type in the globular cluster Messier 5 as a standard timekeeper.[11]   Many of the variables in this cluster rise by a magnitude from minimum to maximum in about 30 minutes, more than doubling the intensity of light emission at minimum.

Professor Bailey's studies of Messier 5 revealed more than 90 cluster-type variables, with periods ranging from 6 to 20 hours and an average period of 13 hours.[12]   The periods of most of these variables are known to within a fraction of a second, according to Bailey's work and the general revision undertaken by the writer with the assistance of Miss Roper.[13]

For the revision of the periods and for the test of the speed of light I made five special series of photographs of Messier 5 in 1917, using the 60-inch reflector of the Mount Wilson Observatory.   The exposures of 20 to 30 minutes that were necessary to record the yellow light with an isochromatic plate and yellow filter were interrupted in the middle for an exposure of 1 or 2 minutes on an ordinary plate sensitive to blue light.   In this manner the variable stars were photographed in two regions of the spectrum at essentially identical times   Photographic and photovisual observations were carried on throughout several hours of the night.   Each run of plates gives fragments of the light curves of all the variables; but since the average period is 13 hours, for only a few stars in each series was the light rising from minimum through median to maximum during a given night's run on the cluster.

The measurement of the plates yielded 6,300 magnitudes, from which 14 measures of the times of median magnitude both in blue and in yellow light were obtained for 11 different variables.   The results appear in Table VIII, II.   The maximum effective wave length for blue light is approximately 4,500

[11] See Appendix C, Refs. 62, 65, 66, 67.
[12] See Section 16; H. A., **78**, Part 2, 1917.
[13] H. B. 851, 1927.

A, and for yellow light, 5,500 A. The observed difference in
the times t of rise to median magnitude

$$\Delta t = t_{pg} - t_{pv}$$

is given in thousandths of a day in the fourth column. Thus,
a positive residual would indicate that the yellow light is measured as arriving first.

TABLE VIII, II—DIFFERENCES IN TIMES OF MEDIAN MAGNITUDES

Number of Variable	Photographic Range	Photovisual Range	Δt	Weight
			d	
1	1 2	0 7	+0 009	3
			—0 001	2
4	1 4	0 9	—0.005	1
8	1 1	0 7	—0 012	1
12	1 3	1 0	—0 005	2
18	1 5	1 05	+0 001	1
20	0 9	0 7	0 000	2
			+0 003	1
28	1 2	0 8	+0 006	2
59	1 0	0 7	—0 003	1
63	1 2	0 9	—0 002	3
64	1 0	0 7	+0 005	1
81	1 1	0 8	+0 001	3
			—0 008	2

It is seen immediately that there is no measurable difference
in velocity, the values of Δt being of the order of the uncertainties of measurement; six are positive, seven are negative,
and one is zero. The mean value of the difference in time
required for the passage of blue and of yellow light over the
distance from the cluster to the earth is

$$\text{Blue} - \text{yellow} = -0^d.00012 \pm 0^d.0007$$
$$= -10 \text{ seconds} \pm 60 \text{ seconds}$$

We have determined in this experiment merely an upper limit
to the difference in speed. We find that since the distance to
the cluster is approximately 11,000 parsecs, light of these two
colors, which differ by 25 per cent in wave length, differs in

time of arrival at the earth by no more than one minute after traversing space for more than 35,000 years. In other words, the relative size of the probable error indicates that the chances are even that the speeds of blue and yellow light do not differ by more than 1 part in 20,000,000,000; probably they do not differ at all.

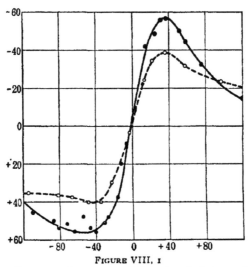

FIGURE VIII, 1

Composite photographic (full line) and photovisual magnitude curves for cluster-type variables in Messier 5   Ordinates are magnitudes in hundredths; abscissae are phases in thousandths of a day.

From 11 determinations of the maxima for 9 variables in Messier 5, a similar equality in speed was found for blue and yellow light. In this result, also, the probable error exceeds the observed average difference, but the determination has much less weight than the one based on median magnitudes.

It might be argued that the simultaneous arrival of the blue and yellow light signals is a coincidence and not a positive indication that the speeds are the same—the delay for one average wave length with respect to another being an integral multiple of the interval between successive maxima or successive

median magnitudes.   Although this argument might have some
weight for a single star, it can be immediately rejected when
several variable stars, at the same distance from the observer
but with slightly differing periods, are involved in the test.

A graphical test of the preceding results is obtained by reducing the light curves of the variables, listed in Table VIII, II,
to two composite curves, one for blue and one for yellow light.
The curves in Figure VIII, 1 were made by reducing all the
variables to a mean period and a mean magnitude range.   For
a given variable v, a suitable linear transformation of the light
intensities is effected by the equations

$$\Delta m = -2.5 \log (1 - \Delta l_0 \Delta l' / \Delta l'_0)$$
and
$$l = 1 - 10^{-0.4\Delta m}$$

in which Δm and Δl are the increments of magnitude and of
intensity measured from maximum, and the prime refers to
minimum light.   For the reduction of the phases t to those
appropriate to a mean period P_0, we have

$$t_v = t_0 - P_v/P_0$$

The coincidence of the photovisual and photographic curves
in Figure VIII, 1 at median and also at maximum magnitude
again illustrates the equality in speed of light in the two colors
But the essential equality in speed does not necessarily indicate
a perfectly transparent interstellar medium.   As Groosmuller,[14]
among others, has pointed out, a large amount of dust and gas
can exist in space without measurably affecting the speed of
light.

[14] Hemel en Dampkring, 22, 153, 1924.

CHAPTER IX

THE TRANSPARENCY OF SPACE

In all researches on the structure of the stellar universe the possibility of the loss of light in its passage through interstellar space must be recognized. The considerable space effect on the colors of stars deduced 15 years ago by Kapteyn and others, if it were finally verified, would be a most serious limitation in the survey of distant clusters and nebulae. Indeed, only for the sun's neighbors could we derive directly the true distances, colors, or temperatures. A spurious redness would invalidate our results and all stars would appear dimmed, the farther away the more affected. Nine magnitudes would be lost at the distance now assigned to the galactic center, and outside galaxies could not be seen at all. The sidereal universe would be effectively "closed" within a relatively short radius, not because of some fundamental property of space-time, but because of the nearly complete opacity of the regions within 100 kiloparsecs.

Fifteen years ago, the possibility of this restricted horizon of a few hundred thousand trillion miles or so brought little dismay to astronomer or philosopher; but now that we have reached some 50,000,000 parsecs and have tasted of the riches that lie beyond our Galaxy, such cramping would be highly irritating.

Another illustration is afforded by the fact that if the loss of the visual light due to absorption or scattering in space should be as much as a millionth of one per cent in each 100,000,000 miles, stars 3,500 light years away would appear about two magnitudes too faint. Uncertainty in the coefficient of scattering is, therefore, very serious in studies of the distances of faint stars, especially if the coefficient is as large as that just suggested.

It is commonly assumed that any dimming that occurs will act through Rayleigh scattering and vary as the inverse fourth power of the wave length of the light. As the effect for blue light is about double that for yellow, the colors of faint and distant stars furnish an obvious method for the detection and measurement of differential scattering.

In addition to the possibility of differential scattering we must consider the blocking of light by meteoric particles or by other means—a phenomenon not directly revealed by star colors but known to occur in extensive Milky Way regions occupied by dark nebulosity. In the following pages we discuss mainly the color tests.

38. Early Investigations of Light Scattering.—At the beginning of the work on star clusters at Mount Wilson (1915) close attention was given to the problem of the scattering and obstruction of light in space.[1] As recently as 1914 Kapteyn[2] had derived from star colors the coefficient,[3] expressed in stellar magnitudes per parsec,

$$d = 0.0003$$

supporting his earlier findings of 1909 and corresponding to a change of a tenth of a magnitude in color index for each 300 parsecs of distance. Among many others, Seeliger, Turner, Kienle,[4] King,[5] Jones,[6] and van Rhijn[7] have considered the question, and the last three, in 1915, had just derived positive values of the absorption coefficient from a consideration of the colors, distances, and proper motions of the available nearby stars.[8]

[1] Mt. W. Contr. 116, 1915.
[2] *Ibid.* 83, 1914.
[3] Unit of distance is one parsec.
[4] For a full bibliography see Kienle's paper in the Jahrbuch der Radioaktivität und Elektronik, 20, 6–9, 1923.
[5] H. A., 59, 182, 1912; 76, 1, 1916.
[6] M. N. R. A. S., 75, 4, 1914.
[7] Dissertation, Groningen, 1915.
[8] Shapley, Mt. W. Comm. 18, 1916.

The following values were then on record for the increase of color index with each parsec of distance:

Observers	Jones	Kapteyn	King	Turner	van Rhijn
Coefficient	0 00047	0 00031	0 0019	0 0030	0.00015

Assuming tacitly that his value of the absorption coefficient was a constant throughout space, Kapteyn proposed optimistically that the excess of color index for distant stars over the values found for nearby stars with similar spectra could be used to measure the distances.   Such scattering effects, however, would doubtless depend on galactic coordinates (especially on latitude) and on the density of nebulosity in the fields through which the light travels.

The test for differential light scattering, based on colors of nearby stars, consists usually in the correlation with proper motion of the excess of redness over the average for a given spectral class, making allowance for other possible factors. Thus, a set of equations of the form

$$\text{C. I.} = a + bm + cp + dM$$

can be solved for the effects on the color index of (1) errors in the scale of apparent magnitude m; (2) scattering of light in space (which should increase with decreasing parallax p); and (3) change of color with absolute magnitude M.   But since

$$M = f(m, \log p)$$

it appears that an unambiguous solution for the constants a, b, c, d is very difficult.   Luminosity and distance effects are not easily differentiated.   Moreover, the magnitudes, parallaxes, and luminosities near the sun are not well known over any considerable range of distance.

In all work yielding positive results, the stars of small proper motion have appeared to be redder.   If the proper motion were taken to be an infallible measure of distance, this excess of redness might mean a scattering of light.   On the other hand, by recognizing relatively small proper motions as a characteristic of highly luminous stars (a relation now firmly established), we see that the excess of redness becomes merely a correlate of bright absolute magnitude. This alternative is now regarded as the correct explanation of the observations that were earlier interpreted as revealing light scattering in the solar neighborhood; the work of Seares and others has clearly shown differences in color for giants and dwarfs.

There remains, however, some evidence of a tenuous localized medium around the sun, which may account in part for the earlier indications of a positive scattering coefficient. King has recently reconsidered the question on the basis of new and accurate magnitudes and colors for the bright stars.[9] He finds support for the thesis that a local cloud of absorbing matter extends to a distance of at least 30 parsecs, reddening the stars at the rate of $d = 0.0003$ magnitudes per parsec but affecting more distant stars only by a constant (and negligible) amount. Similar localized effects have been found for nebulous stars, and perhaps, also, in some galactic clusters; but the whole color discrepancy measured by King is extremely small and uncertain in amount, since the base line is relatively short.

If instead of being confined to stars in our immediate stellar system the study of colors is extended to much more distant objects, it can readily be decided whether general light scattering is to play an important part in stellar investigations. We can, fortunately, take advantage of great distances in the present study of globular clusters.

39. Blue Stars in Messier 13.—After Kapteyn's work on light scattering, and the contemporary results of King, Jones, and van Rhijn, which independently confirmed it, the discovery in 1915 of stars with negative color indices in Messier 13 was unexpected. A critical examination of photovisual and photographic magnitude scales failed to assign the apparent blueness of the cluster stars to observational error. Out of 495 stars with well-determined color indices, 86 were found to be of color class b, and 63 of class a.

[9] H. C. 299, 1927.

With a scattering coefficient of 0.0003 (Kapteyn's last value) and an assumed distance for the cluster of as little as 1,000 parsecs, practically no negative colors should appear if the stars are of usual spectral classes. With the obviously better parallax of 0″.0001, the color index produced by Kapteyn scattering

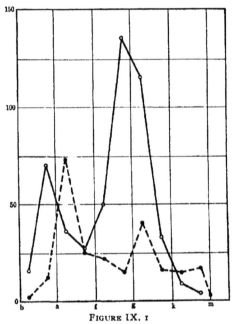

FIGURE IX, 1

Frequency of color classes in Messier 13 (full line) and in the neighborhood of the sun (Yerkes actinometry). Coordinates are relative numbers of stars and color classes.

should be +3.0 for stars of spectral class A, and many color indices should be greater than +4 in a typical distribution of spectral classes. (The direct correspondence of color class with spectral class in this cluster had been provisionally verified by the spectrograms made by Pease.[10])

It is found that the range of color index observed in Messier 13 is the same as that found among nearby galactic stars, as

[10] See Chapter III, Section 12.

illustrated in Figure IX, 1. There are no color indices in the cluster larger than $+2$. The largest admissible color excess in the cluster appears to be about $0^m.1$; and, even if this excess is attributed entirely to light scattering, we derive $d = 0.00001$, a thirtieth of the value derived by Kapteyn and a fifteenth of the value found by van Rhijn from a consideration of the stars in the Yerkes Actinometry. The foregoing upper limit of scattering is so small as to be entirely negligible in dealing with nearby stars; but a more accurate value is desirable, and fortunately it can be found from studies of external stellar systems.

40. Faint Blue Stars in the Milky Way.—The galactic latitude of Messier 13 is $+40°$. That light scattering is absent in this direction is no proof that it does not occur elsewhere, especially in low galactic latitudes where stars and diffuse nebulosity are concentrated. To make the test more comprehensive with respect to galactic latitude and longitude, the search for negative color indices has been carried out on several distant cluster systems, on the assumption that a normal range of color index and the presence of numerous blue stars are sufficient evidence of the transparency of space in the directions and to the distances considered. The results are in Table IX, I, where the observed extremes of color are shown in the fourth and fifth columns, and, in the next two columns, the mean photovisual magnitude and mean color index for a group (usually ten) of the faintest blue stars. The distances in the last column are taken from the tables in Appendices A and B. Similar results have been obtained for faint stars in four Milky Way fields,[11] and Seares has found faint blue stars in the Selected Areas that fall in low galactic latitude.

Considering the relatively large distances of the tabulated objects, and their wide distribution, we appear justified in generalizing the results of the study of Messier 13 and in concluding that interstellar media, throughout the distances here concerned, have no serious effect on the color of light. This

[11] Shapley, Mt. W. Comm. 44, 1917.

conclusion does not bear on obstruction of light by recognized diffuse nebulosity (dark and bright) or on the colors of nebulous stars.[12]

TABLE IX, I.—Test for Space Transparency in Various Regions

Cluster	Galactic		Color Index		Mean Pv. Mag	Mean Color Index	Distance
	Long	Lat	Largest	Smallest			
	°	°					Kpc.
Messier 3	12	+78	+1 77	−0 39	15 1	−0 16	12 2
13	27	+40	+1 42	−0 52	16 54	−0 34	10 3
15	33	−28	+1 50	−0 21	16 0	−0.14	13 1
38	139	+ 1	+2 12	−0 45	13 5	−0.16	1 1
36	142	+ 1	+1 50	−0 30	12 5	−0 23	1 2
35	154	+ 3	+1 31	−0 15	11 5	−0 07	0 8
50	189	− 1	+2 00	−0 04	12 3	−0 02	0 8·
5	332	+46	+1 67	−0 11	14 6	−0 10	10 8
22	338	− 9	+2 05	−0 45	14 34	−0 28	6 8
11	355	− 3	+2 06	−0 16	14 32	−0 08	1 2

41. N.G.C. 7006 and the Scattering of Light.—Notwithstanding the great distance and consequent faintness of N. G. C. 7006, the magnitudes and color indices of 38 of its supergiant red stars have been measured.[13] The distribution of color in this cluster has already been compared in Chapter III with that of the brighter stars in Messier 13 and Messier 3. Even at a distance more than five times that of Messier 13, no abnormal redness appears in N. G. C. 7006; and there is no other peculiarity in its star colors, although its radiation has traveled through the scattering materials of space for an interval of more than 180,000 years.

Comparing N. G. C. 7006 with Messier 3 and Messier 13, we have, as mean color indices for the brightest 35 stars:

N G. C. 7006   1 10
Messier   3   1.15
Messier 13   1 03

[12] Seares and Hubble, Mt. W Contr. 187, 1920.
[13] Shapley, Mt. W. Contr. 156, 5, 1918; see, also, Shapley and Mayberry, Mt. W. Comm. 74, 1921.

If interstellar media have any effect at all on the color indices of stars in this distant system, the reddening apparently does not exceed a tenth of a magnitude. The absorption coefficient, therefore, expressed as change of color index for each parsec of distance, is

$$d < 0.000002$$

42. The Coma-Virgo Group of Nebulae.—With a preliminary estimate of the distance of the Andromeda Nebula, Lundmark and Lindblad found[14] for the coefficient of space absorption

$$d < 0.000002$$

They used the method of effective wave lengths as an indicator of color and obtained for Messier 31 and for fainter nebulae values which were not appreciably different from those for average stars.

A subsequent investigation of the integrated magnitudes and colors in the cloud of 300 bright extra-galactic nebulae in Coma and Virgo[15] permits a further consideration of light scattering. The photographic magnitudes are derived from Harvard plates. The visual magnitudes are from Holetscheck, reduced to the Harvard scale by way of the corrections to the faint magnitudes of the Bonn Durchmusterung, as evaluated by Pickering and Pannekoek. The mean color index for 41 nebulae fainter than photographic magnitude 11.5 is +0.69 ±0.04, in good agreement with spectral class G, which is normal for nebulae of this class. The color excess, if existent, is, therefore, not likely to be greater than two tenths of a magnitude. The distance of the cloud of nebulae is of the order of 3×10^6 parsecs, and therefore we find[16]

$$d < 0.0000007$$

[14] Ap. J., 50, 386, 1919. Through an error the value is printed ten times too small.

[15] Shapley and Ames, H. C. 294, 1926.

[16] The value in H. C. 294 is erroneously printed ten times too large. See H. B. 841, 1926.

This value is the color effect alone; the total loss photographically is about double the color-index change.

Apparently, we need not disturb ourselves further about the general dimming of light in space, even when dealing with external stellar systems, unless it happens that the diminution differs from molecular scattering and has no effect on colors.

43. The Obstruction of Light in Space.—Dark nebulosities effectively conceal at least one or two per cent of the whole sky. Because of their presence in the Sagittarius region, they hide a much higher percentage of the total number of galactic stars. The question for consideration here is the extent to which such diffuse nebulous material in the sky at large obstructs starlight without affecting colors. That the effect is inappreciable, outside the regions of recognized nebulosity, has in the past been inferred rather than proved. The inference is based on (1) the practically complete absence of differential light scattering, indicating the meagerness of dust particles of small dimensions and presumably, therefore, of large dimensions; and (2) the fact that the angular dimensions of globular clusters decrease systematically with decreasing apparent brightness. The decrease of cluster diameter with brightness also holds true for the magnitudes of the individual stars in globular clusters.

The theoretical relation in transparent space, $d = 10^{-0.2m + \text{const}}$, is approximately maintained for globular clusters,[17] but the agreement with theory has little meaning because neither the available integrated magnitudes m nor the catalogued angular diameters d are on true scales representing luminosity and linear dimensions. As far as it goes, the result for the globular clusters indicates the essential transparency of space up to a distance of 100,000 light years. The serious observational difficulties in measuring the true angular diameters[18] and total magnitudes, and the dispersion in linear dimensions leave us,

[17] Shapley, H. B. 864, 1929.
[18] See Chapter XIV, Section 72

however, uncertain concerning general light obstruction near the
Galaxy, even though differential light scattering is inappreciable.

The uncertainties that attend a quantitative test of obstruction by means of globular clusters are largely avoided by the
use of extra-galactic nebulae. The boundaries of at least some
of these objects are more clearly defined on photographic plates

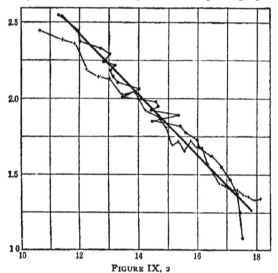

FIGURE IX, 2

The relation of angular diameter to apparent photographic
brightness for 2,750 nebulae in the Coma-Virgo region.
Ordinates are logarithms of diameter; abscissae are total
magnitudes. (See H. B. 864.)

than are those of clusters, for which the limits are indiscernible
because of the confusion with foreground and background stars
and the irregularities in adjacent stellar distribution. Figure
IX, 2 shows the test of the transparency of space in the Coma-
Virgo region between the relatively bright Cloud A and the
clouds B, C, and D whose members average from three to
six magnitudes fainter. The magnitudes are photographic,
and the diameters are expressed in seconds of arc.[19] The
two broken lines in the figure connect the logarithms of the

[19] Shapley and Ames, H. B. 864, 1929.

means of the diameters for intervals of magnitudes (crosses) and the mean magnitudes for intervals of angular diameters (dots). The straight line gives the theoretical relation for transparent space, $m = 24.15 + 5 \log d$. The magnitude 24.15, for which the average galaxy here considered would have an angular diameter of $1''$, has, according to the diagram, an uncertainty of a tenth of a magnitude. The observed value of the constant would vary somewhat, of course, with the kind of photographic plate and the method of measuring diameters.

Because of the possibility that the measured magnitude and diameter of extra-galactic nebulae would be equally affected by light obstruction, the foregoing test needs further consideration. The nebulae in the Coma-Virgo region have been placed in six grades, from a to f, in order of increasing central concentration.[20] Grade a represents an essentially uniform surface brightness; for it an obstructing medium should affect total magnitudes without disturbing angular diameters, and consequently the theoretical relation would fail.

It is improbable that if there is an appreciable obstruction of light in space the relation of diameter to apparent magnitude would be the same for all grades. Those having marked central concentration should exhibit a more rapid decrease in diameter with decreasing apparent magnitude than is shown by the grades of more even surface brightness. But this change of rate of decrease is not observed in Figure IX, 3, which is the same as the preceding figure except that the various grades are plotted separately.

From the representation of the observations by the straight lines in Figures IX, 2 and IX, 3 we conclude that space, at least in the direction of Coma-Virgo, is effectively transparent throughout a distance of more than 100,000,000 light years. We may allow, tentatively, two-tenths of a magnitude for the total obstruction, corresponding to a loss in light, expressed in magnitudes per parsec, of

$$l < 0.000000007$$

[20] *Ibid* , H. B. 862, p. 20, 1929; see, also, H. B. 869, p. 32, 1929.

This limiting value of loss through obstruction should be considered as an indication of the size of the observational uncertainties rather than as a positive measure of the amount of absorbing material of all kinds in space. It depends on a direct test and is about twenty times smaller than the earlier value, which depended on inference. It is of higher weight

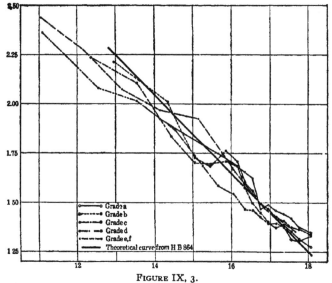

FIGURE IX, 3.

The relation of angular diameter to apparent photographic brightness for nebulae in the Coma-Virgo region, plotted according to grades of central concentration. (See H. B. 866.)

because of the much greater amount of material involved and should give us confidence in our explorations of remote parts of space.

Although we have shown that scattering and obstruction of light in many directions in high galactic latitudes are probably negligible, there is growing evidence of some general absorption or scattering in the direction of a few of the Milky Way star clouds—an effect amounting to two or three tenths of a magnitude.

CHAPTER X

THE PERIOD-LUMINOSITY CURVE

WE do not certainly know what a Cepheid variable star is or how the variability arises or dies away. We have vast accumulations of observations, but still too few; we have numerous theories and interpretations, probably too many; and we remain in the dark concerning some of the most fundamental properties of the Cepheid. Nevertheless, in spite of our ignorance concerning them, the Cepheids are accepted as the most useful type of star in the sky. Luckily, they are widely spread—in the solar neighborhood, in the star clouds of the Milky Way, in clusters, in the Magellanic Clouds, and in external galaxies. Their spectral and temperature changes continually raise questions about the age and evolution of supergiant stars. But by far their most intriguing characteristic is the property of revealing candle power and distance simply through the period of the variation in light. It is this property of the Cepheids, empirically found, developed, and used, that in the past 15 years has opened up the Galaxy and many remote systems that lie beyond.

44. Historical Notes.—The relation between the absolute magnitude of a Cepheid variable and the logarithm of its period, which I have called the "period-luminosity relation," was first presented empirically in 1912 by Miss Leavitt in a brief study of some of the variable stars in the Small Magellanic Cloud. She dealt, however, with only the apparent magnitudes of 25 stars with periods in excess of $1^d.2$ and did not consider matters of absolute magnitude, distances, and the relation to cluster-type variables. The high absolute luminosity of a few Cepheid variables had already been inferred by Hertzsprung[1] from a

---

[1] Zeit. f. Wiss. Phot., 5, 94, 107, 1907.

consideration of their proper motions; and later he estimated the distances and plotted the distribution of 45 galactic Cepheids,[2] using Miss Leavitt's provisional linear relation between magnitude and the logarithm of the period, and for the zero point deriving a mean luminosity of typical Cepheids from Boss' proper motions. Russell also noted the high luminosity of the Cepheids in Boss' catalogue.[3]

Practically all subsequent work on the period-luminosity curve, and its development for use in the measurement of distances, has been done as an incidental but important part of the systematic investigation of clusters at Mount Wilson and later at Harvard.[4] There have also been some serious attempts at theoretical interpretations of the period-luminosity relation, including those of Eddington,[5] Seares,[6] and Shapley,[7] and some vigorous critical discussions of the data on galactic Cepheids.[8]

45. Miss Leavitt's Work on the Periods of Cepheids.— In her first discussion of the magnitude estimates of the brighter Cepheids in the Small Magellanic Cloud, Miss Leavitt gave the provisional periods of 16 variables and noted that the brighter stars had the longer periods.[9] In her later communication[10] the number was increased to 25, with a wide range in period and brightness. The essential data are reproduced in Table X, I, in which successive columns contain the Harvard number of the variable, its period, and the maximum and minimum magnitudes on a provisional scale.

Miss Leavitt's plot of the periods and of maximum and minimum magnitudes is reproduced in Figures X, 1 and X, 2.

[2] A. N., **196**, 201, 1913.

[3] Science, N. S., **37**, 651, 1913.

[4] Shapley, Mt. W. Contr. 151, 1917; H. C. 280, 1925.

[5] M. N. R. A. S., **79**, 2, 1918.

[6] Mt. W. Contr. 226, 40–56, 1921.

[7] *Ibid.* 154, 1917; 190, 7, 1920, H C. 314, 1927; 315, 1927.

[8] Kapteyn and van Rhijn, B. A. N., No. 8, 1922; Curtis, Bul. Nat. Res. Coun., **2**, 194, 1921; R. E. Wilson, A. J., **35**, 35, 1923.

[9] H. A , **60**, 105, 1908.

[10] H. C. 173, 1912.

Apparently, neither Miss Leavitt nor Pickering felt certain at that time that the Magellanic Cloud variables were of the same type as galactic Cepheids.   Miss Leavitt wrote:

TABLE X, I.—MISS LEAVITT'S OBSERVATIONS OF CEPHEIDS IN THE SMALL MAGELLANIC CLOUD

H V.	Period	Maximum	Minimum	H. V	Period	Maximum	Minimum
	d	m	m		d	m	m
1505	1 25336	14 8	16 1	1400	6 650	14 1	14 8
1436	1 6637	14 8	16 4	1355	7 483	14 0	14 8
1446	1 7620	14 8	16 4	1374	8 397	13 9	15 2
1506	1 87502	15 1	16 3	818	10 336	13 6	14 7
1413	2 17352	14 7	15 6	1610	11 645	13 4	14 6
1460	2 913	14 4	15 7	1365	12 417	13 8	14 8
1422	3 501	14 7	15 9	1351	13 08	13 4	14 4
842	4 2897	14 6	16 1	827	13 47	13 4	14 3
1425	4 547	14 3	15 3	822	16 75	13 0	14 6
1742	4 9866	14 3	15 5	823	31 94	12 2	14 1
1646	5 311	14 4	15 4	824	65 8	11 4	12 8
1649	5 323	14 3	15 2	821	127 9	11 2	12 1
1492	6 2926	13 8	14 8				

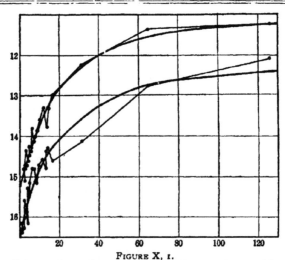

FIGURE X, I.

Relation of period to maximum and minimum photographic magnitudes for variables in the Small Magellanic Cloud. Ordinates, magnitudes; abscissae, periods in days.

They resemble the variables found in globular clusters, diminishing slowly in brightness, remaining near minimum for the greater part of the time, and increasing very rapidly to a brief maximum.

But that the importance of the observed relation did not escape Miss Leavitt is shown by her remarks:

Since the variables are probably at nearly the same distance from the earth, their periods are apparently associated with their actual emission of light, as determined by their mass, density, and surface brightness . . . Two fundamental questions upon which light may be thrown [by a study

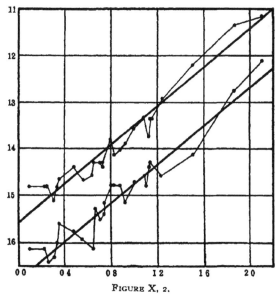

FIGURE X, 2.

Relation of period to maximum and minimum magnitudes for variables in the Small Magellanic Cloud. Coordinates are photographic magnitudes and logarithms of the periods.

of the Magellanic variables] are whether there are definite limits to the mass of variable stars of the cluster type, and if the spectra of such variables having long periods differ from those whose periods are short.

46. The Visual Period-luminosity Curve.—It will serve no useful purpose to recount in detail the development of the period-luminosity relation; the steps are given fully in the various papers referred to in the bibliography, especially in Mount Wilson Contribution 151, published in 1917. It will suffice to reproduce in Figure X, 3 the original visual period-luminosity curve and give a tabulation of the smooth curve in Table X, II.   In reproducing the table and curve, an arbitrary correction of $+o^m.23$, described below in Section 52, has been applied to the absolute magnitudes.   The plotted points in the figure refer to the various clusters and the Small Magellanic

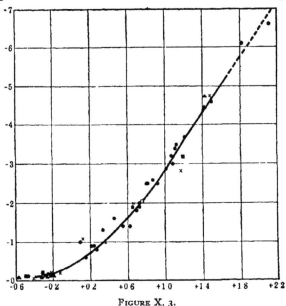

FIGURE X, 3.

The visual period-luminosity curve   Coordinates are absolute visual magnitudes and logarithms of the periods   The curve is made up from five systems the Small Magellanic Cloud (dots), ω Centauri (crosses), Messier 5 (triangles), Messier 3 (squares), and Messier 15 (circled crosses)   Most of the symbols for periods less than a day represent averages of about ten variables.   The data are taken from Mount Wilson Contribution 151, 1918; variable 50 in Messier 5 is omitted.   The line represents the period-luminosity curve

Cloud, from which the data on the Cepheids were derived. From this plot I have now removed the points referring to galactic Cepheids, because their relative distances are not accurately and independently known.

The observations on which the visual period-luminosity curve was based were mainly photographic and were reduced to the visual system by means of a provisional period-color-index relation based on the observations of spectra available in 1917. From the determination of the period-spectrum relation, given later in this chapter, it has been shown that the original reduction from photographic to visual magnitudes was very rough, as was suspected at the time. Hence, the visual period-luminosity curve here given should not be used without correction with a more accurate reduction from photographic to visual magnitude.

TABLE X, II.—COORDINATES OF THE VISUAL PERIOD-LUMINOSITY CURVE

Logarithm of Period	Absolute Visual Magnitude	Logarithm of Period	Absolute Visual Magnitude
—0 6	—0 11	+0 8	—2 20
—0 4	—0 10	+1 0	—2 92
—0 2	—0 15	+1 2	—3 64
0 0	—0 41	+1 4	—4 36
+0 2	—0 76	+1 6	—5 08
+0 4	—1 14	+1 8	—5 79
+0 6	—1 58	+2 0	—6 5·

It should be noted especially that high visual absolute magnitudes of Cepheids with periods in excess of 30 days are reduced by at least half a magnitude by the later studies of the relation of period to spectrum and color. But even with these modifications, the absolute magnitudes of the longest-period Cepheids in the globular clusters and in the Magellanic Clouds remain among the highest known. Only some of the novae and the brightest supergiant stars of the Magellanic Clouds appear to excel them in luminosity.

47. The Periods of 106 Variables in the Small Magellanic Cloud.—Since the practical use of the period-luminosity relation soon became confined almost exclusively to photographic work, it seemed advisable to set up a well-determined photographic period-luminosity curve for the study of the distances of clusters, star clouds, and spiral nebulae. Granting the generality of the relation, we deemed it best, moreover, to base the curve on material from a single system.   Hence, we have made an effort to fix the form of the curve as accurately as possible through a comprehensive study of the variable stars in the Small Magellanic Cloud.   The faintness of the variables and the probable difficulties with the Eberhard effect and with occasional wisps of obscuring nebulosity may produce large accidental errors for the individual stars; but the number of stars compensates for individual deviations.

The establishment of dependable magnitudes throughout the Small Magellanic Cloud has been a laborious preliminary to the study of the variable stars.   In all, photographic magnitudes have been determined in the Cloud for 25 sequences, involving a total of approximately 400 stars.   The brightest magnitude is 6.73, and the faintest in most of the sequences is between 17 and 18.   The magnitude scale is based on the present International North Polar Standards, having been connected through the Harvard Standard Regions C_1 and C_2, in declination $-15°$ and right ascensions 1^h and 3^h.[11]   It is believed that the average deviation of the scale for sequences in the Small Cloud is not over two per cent, throughout the interval from the tenth to the seventeenth magnitudes, where the Cepheid variables under discussion are found.   Uncertainties still remaining in the zero point of the magnitude scale, at least for some of the secondary sequences, may be of the order of two-tenths of a magnitude.

For 32 variables in the Small Cloud, periods were determined by Miss Leavitt.[12]   For 74 the periods were originally determined by Yamamoto, working with Harvard plates, and later many were revised and checked by Miss Wilson or the writer.[13]

The 106 variables of the Cepheid class are well scattered throughout the Cloud, and the magnitudes are referred to

[11] Shapley, H. C. 255, 1924.
[12] Twenty-five were published by Miss Leavitt in H. C. 173, 1912, and the others by Shapley, Yamamoto, and Wilson (*loc. cit., infra*).
[13] H. C. 280, 1925.

many different sequences. In one field, in the densest part of the Cloud, some of the variables are at least half a magnitude fainter than is normal for their period. This discrepancy may be due to the Eberhard effect or to an effect of uneven background. Except for these few stars, however, the relation of median magnitude to period appears to be the same throughout the whole of the Small Cloud; but in the Large Cloud the current study of the variable stars indicates more frequent disturbance of magnitudes by the extensive nebulosities.

48. The Photographic Period-luminosity Curve.—The plot of the median apparent photographic magnitudes against logarithms of the periods for the 106 individual variables is shown in Figure X, 4. Points of low weight are enclosed in circles, and the periods determined by Miss Leavitt are indicated by crosses. A few of the stars diverging most widely from the curve may not be actual members of the system, but the fairly high galactic latitude ($-33°$) makes rather unlikely the occurrence of typical Cepheids except as members of the Cloud.

The deviations from the curve in Figure X, 4 are not larger than might be expected in view of the difficulties of observation. The average deviation in magnitude of the 32 Leavitt variables is 0.19; for the 74 stars it is 0.25; for all, 0.23. The systematic magnitude deviation from the curve is -0.063 for the Leavitt variables and $+0.049$ for the others, showing a negligible systematic difference between the earlier and the later work.

Some of the dispersion in magnitude can be attributed to the thickness of the Cloud in the line of sight, but between the variables at the near and far edges this correction amounts to only 0.14 magnitudes. If α is the angular radius of a sensibly spherical system, the maximum dispersion in magnitude arising from the thickness is given with sufficient approximation for remote systems by

$$\Delta m = 5 \log \left[\frac{(1 + \sin \alpha)}{(1 - \sin \alpha)} \right]$$

The correction is thus independent of the distance and real thickness.   The diameter of the Small Cloud is 3°.6, and the extreme correction to mean values is, therefore, less than a tenth of a magnitude

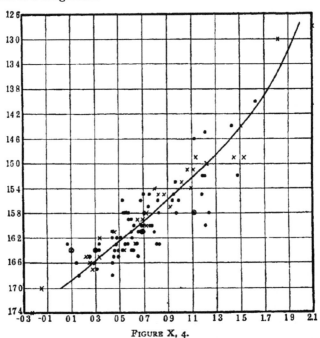

FIGURE X, 4.

Photographic period-luminosity curve. Ordinates are apparent magnitudes; abscissae are logarithms of the periods. Crosses refer to periods determined by Miss Leavitt. Circles enclose values of low weight.

As noted above, other factors contributing to residuals from the period-luminosity curve are the Eberhard effect and obstruction by nebulosity. To these may be added actual deviations of the periods and magnitudes from average conditions— that is, a true scattering which may arise from differences in mass, structure, or other physical properties. We should also consider errors in the periods and the observational uncertainties for individual variables, which may easily amount to two tenths of a magnitude. Errors due to the failure to resolve double stars are not likely to be serious.

In Figure X, 4 only two stars with periods shorter than a day are included. There appear to be many others of this sort in the Cloud,[14] but they are not considered in the present discussion because of the relative weakness of the magnitude scale fainter than 17.0. These cluster-type variables will soon be studied with a large reflector, when short-exposure photographs and better magnitude sequences can be used.

By grouping the points plotted in Figure X, 4, first in order of the logarithms of the period and then in order of apparent magnitude, we get the two sets of means shown in Table X, III. All magnitudes of Cepheids in this chapter are median values: (max. + min.)/2.

TABLE X, III.—DATA FOR THE PHOTOGRAPHIC PERIOD-LUMINOSITY CURVE

Mean Log P	Mean Magnitude	Number	Absolute Magnitude	Mean Magnitude	Mean Log P	Number	Absolute Magnitude
−0 194	17 2	2	−0 12	17 00	+0 054	4	−0 32
+0 253	16 51	17	−0 81	16 44	0 437	33	−0 88
0 610	16.08	53	−1 24	15 99	0 677	36	−1 33
0 986	15.48	21	−1 84	15 51	0 871	17	−1 81
1 357	14 93	11	−2 39	15 05	1 267	9	−2 27
1 961	12 90	2	−4 42	14 38	1 380	5	−2 94
				12 90	1 961	2	−4 42

The distance modulus for the Small Magellanic Cloud (see Chapter XIII) is 17.32, and the absolute magnitudes for variables in the Cloud are therefore given by $M = m - 17.32$. In the fourth and eighth columns of Table X, III are the absolute magnitudes corresponding to mean values of log P. The plot of these quantities in Figure X, 5 gives the adopted photographic period-luminosity curve, which is extended to log $P = -0.6$ on the basis of the data embodied in the visual period-luminosity curve, Figure X, 3. The curve is tabulated for equal intervals of log P in Table X, IV. An adjustment of the lower end of the curve may result from studies of the faintest variables in the Magellanic Clouds.

[14] Shapley, P. N. A. S., 8, 69, 1922.

TABLE X, IV.—COORDINATES OF THE PHOTOGRAPHIC PERIOD-LUMINOSITY CURVE

Logarithm of Period	Absolute Photographic Magnitude	Logarithm of Period	Absolute Photographic Magnitude
−0 6	0 00	+0 8	−1 53
−0 4	0 00	+1 0	−1 89
−0 2	−0 07	+1.2	−2 26
0 0	−0 31	+1 4	−2 68
+0 2	−0 61	+1 6	−3.19
+0 4	−0 93	+1 8	−3 81
+0 6	−1 22	+2 0	−4 60

The photographic period-luminosity curve can be shifted as a whole, up or down the absolute magnitude axis, if such a

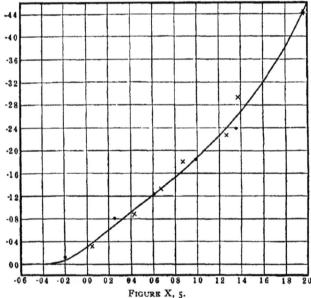

FIGURE X, 5.

Adopted photographic period-luminosity curve. Coordinates are absolute photographic magnitudes and logarithms of the periods Dots represent means for intervals of logarithms of periods; crosses, for intervals of magnitude.

change is justified by future studies of the proper motions and mean absolute magnitudes of galactic Cepheids; but, of course, this revision of the zero point would not affect the general form of the curve.

49. The Period-spectrum Relation.—In my first work on star clusters the need of combining visual observations of galactic Cepheids with photographic observations of the variables in globular clusters and the Magellanic Clouds led to a preliminary attempt to determine the dependence of color index (spectral class) on period for galactic Cepheids.[15] Attempts to improve on the data through direct observations of Cepheid spectra resulted in the discovery that the spectra of all Cepheids are variable—they show a systematic and continuous variation of spectral class, synchronous with the changes in magnitude and velocity, the type at maximum being earlier than the type at minimum.[16]

That the longer period Cepheids frequently show later spectrum classes (or redder color indices) than those of short period has long been known, mainly from early work at the Harvard and Lick observatories; but the data I obtained at Mount Wilson on the changes of the spectra of 20 stars were not sufficient to revise my preliminary period-spectrum curve. The revision has recently become possible through a much more extensive investigation made jointly with Miss Walton, using Harvard plates.[17] The work is described here as a preliminary to the derivation, in the following section, of a period-luminosity curve for galactic Cepheids.

The determination of the median spectra and the spectral variations for 70 Cepheid variable stars is based on the examination of more than 1,200 spectrum plates. The classifications are by Miss Cannon. A discussion of the data brings out the following points:

1. Among cluster-type Cepheids of the galactic system there is a considerable scattering of median spectral class, but appar-

---

[15] Shapley, Mt. W. Contr. 92, 16, 1914.
[16] *Ibid.* 124, 2, 1916; Mt. W. Comm. 14, 21, 22, 27, 1915–1916
[17] H C 313, 1927.

ently no progressive change of spectrum with variation in period.[18]

2. The average deviation from the mean spectrum-period curve for all 70 Cepheids is 2.1 spectrum units.

3. No progressive dependence of spectral range on the length of period, or on the median spectral class, is found. The average range is slightly more than one spectral subdivision.

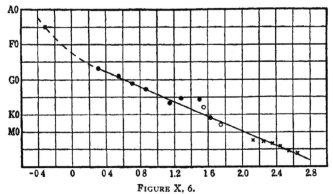

FIGURE X, 6.

The relation of spectral class (ordinates) to logarithm of the period (abscissae) for variable stars Open circles refer to RV Tauri variables, crosses to long-period variables, dots to Cepheid variables, and the open squares to the mean cluster-type Cepheids.

4. In Table X, V the mean spectrum and the logarithm of the corresponding effective temperature are given for intervals of the logarithm of the period (Figure X, 6). The table includes not only data for the cluster-type Cepheids, supplemented by 19 values from the Henry Draper Catalogue, and for classical Cepheids, but also corresponding data for 11 RV Tauri variables, assembled in two groups, and 389 long-period variables (Class Me). The last three entries in the table for classes N, R, and S are provisional, the temperatures not yet safely estimated. Indeed, the tabulated temperatures for long-period variables are none too dependable.

[18] See Chapter IV, where the peculiar properties of cluster-type Cepheids are considered in some detail

TABLE X, V.—RELATION OF SPECTRUM TO PERIOD FOR VARIABLE STARS

Log P	Spectrum	Number	Log T	Log P	Spectrum	Number	Log T
−0 32	A5 0	25	3 93	2 13	M2.4e	26	3 40
+0 31	F7	2	3 78	2 25	M2 9e	27	3 38
0 55	F9.1	8	3 76	2 35	M3.3e	65	3.365
0.71	G1 1	12	3 73	2 45	M4 1e	102	3 335
0 87	G2 9	18	3 71	2 55	M5 4e	113	3 28
1 15	G6 8	6	3 65	2 65	M6.1e	46	3 24
1.29	G5 3	9	3 67				
1.50	G5.7	5	3 66	2 57	N	27	3 42
1.63	K1	2	3 60	2 58	R	5	
				2 56	S	27	
1 56	G8	5	3 64				
1 75	K3	6	3 57				

50. The Period-luminosity Relation among Galactic Cepheids.

—That a period-luminosity relation exists for Cepheid variables in star clouds, in clusters, and in spiral nebulae is generally recognized. But the indication of a definite period-luminosity relation for galactic Cepheids is admittedly not strong. This might be taken as evidence for a greater real dispersion in the properties of galactic Cepheids—a somewhat unlikely supposition.

The main reason for indefiniteness in the period-luminosity relation among nearby Cepheids is obviously our ignorance of the parallaxes and luminosities. There are few Cepheids nearer than 200 parsecs; hence, few precise trigonometric parallaxes are available. The proper motions, likewise, are as yet scarcely sufficient for the derivation of a mean value of the parallax, to say nothing of discriminating between the absolute luminosities for different periods. The trigonometric parallaxes of the Cepheids, as given in the system of the Yale Catalogue, are at present as in Table X, VI. This material, showing most of the parallaxes to be less than their probable errors, strongly supports the deduction from proper motions that the galactic Cepheids are certainly remote and highly luminous.

The trigonometric values are in the mean systematically o".oo1 larger than the values from the period-luminosity relation—a quite negligible difference in view of the observational errors. The individual differences for 10 out of the 16 stars are actually smaller than the probable errors of the corresponding trigonometric values

TABLE X, VI.—TRIGONOMETRIC PARALLAXES OF CEPHEID VARIABLES

Star	Parallax		Source	Per Lum Parallax	Difference Tr − P L
	"	"		"	"
Polaris	o oo7 ± o oo7		G, F+	o o18	−o o11
SU Cas	7	7	Mt. W	4	+ 3
SZ Tau	13	10	M, Mu	2	+ 11
RX Aur	2	7	Mt. W	1	− 3
T Mon	7	9	McC	1	− 8
RT Aur	8	5	WSM	4	+ 4
ζGem	5	6	A, M, S+	4	+ 1
RY Boo	13	8	Mt. W	1	+ 12
RR Lyr	8	4	A, W+	3	+ 5
U Vul	o	10	McC	1	− 1
η Aql	3	11	McC	5	− 2
S Sge	5	8	M, Mu	2	+ 3
X Cyg	3	12	Sprl	1	+ 2
T Vul	19	9	McC	3	+ 16
β Cep	7	7	M, Yk	22	− 15
δ Cep	9	7	A, M	6	+ 3

The direct measures of parallax obviously do not yet suffice as a test for dependence of luminosity on period. The proper motion data compiled by R. E. Wilson[19] are, however, more extensive and useful for the purpose.

All Cepheid variables for which the probable error of the proper motion is less than ten thousandths of a second are entered in order of period in Table X, VII The median visual magnitudes are from my compilation in Mount Wilson Contribution 155 or from more recent Harvard data[20]; the proper motions are from Wilson's paper, and the spectral classes from

[19] A. J., 35, 35, 1923.
[20] Walton, H. B. 845, 1927.

Shapley and Walton[21] or, if italicized, from the Henry Draper Catalogue. In grouping the material of this table for intervals of the logarithm of the period, and for intervals of the reduced

TABLE X, VII.—PROPER MOTIONS OF GALACTIC CEPHEIDS

Star	Median Vis Mag.	Log Period	Total P M	Reduced P M	Spectrum
	m		*"*	*"*	
U Car	7 4	1 59	0 001	0 001	G6 s
l Car	4 3	1 55	0 031	0 012	G7
T Mon	6 2	1 43	0 030	0 021	G4 s
RY Sco	8 2	1 31	0 023	0 040	G3 s
W Vir	9 6	1 24	0 083	0 280	*Pec.*
Y Oph	6 3	1 23	0 003	0 002	G2
X Cyg	6 5	1 21	0 018	0 014	G4 s
TT Aql	7 6	1 14	0 015	0 020	G6
XX Cen	7 4	1 04	0 022	0 026	*G5*
ζ Gem	4 0	1 01	0 010	0 002	*G0p*
S Nor	7 1	0 99	0 009	0 009	*G0p*
β Dor	3 9	0 99	0 015	0 004	F5 s
S Mus	6 8	0 98	0 019	0 021	G1 s
κ Pav	4 5	0 96	0 017	0 005	*F5p*
S Sge	5 8	0 92	0 006	0 004	G3
U Vul	7 0	0 90	0 031	0 031	G4
W Gem	7 1	0 90	0 010	0 010	G0 s
ER Car	7 1	0 89	0 014	0 015	*F8*
W Sgr	4 7	0 88	0 011	0 004	G0 s
R Mus	7 0	0 88	0 008	0 008	G1 s
η Aql	4 0	0 86	0 013	0 003	G4
U Aql	6 6	0 85	0 005	0 004	G3 s
X Sgr	4 7	0 85	0 018	0 006	G3 s
BB Sgr	6 8	0 83	0 049	0 045	G4
U Sgr	6 9	0 83	0 011	0 010	G4
T Cru	7 2	0 83	0 018	0 020	G2 s
V Car	7 8	0 83	0 018	0 026	G5 s
S TrA	6 9	0 80	0 009	0 009	G2 s
RV Sco	7 1	0 78	0 028	0 030	*F5*
R Cru	7 3	0 77	0 017	0 020	G4
Y Sgr	5 8	0 76	0 022	0 013	G0 s
V Cen	7 1	0 74	0 032	0 034	G1
δ Cep	4 1	0 73	0 011	0 003	G2
AP Sgr	7 3	0 70	0 015	0 017	*F5*
S Cru	7 0	0 67	0 044	0 044	G1 s
T Vul	5 8	0 65	0 000	0 000	*F8p*
BF Oph	6 8	0 61	0 003	0 003	*G5*
α UMi	2 1	0 60	0 046	0 005	*F8*
SU Cyg	6 6	0 58	0 024	0 020	F5 s
RT Aur	5 3	0 57	0 024	0 011	F9
R TrA	7 0	0 53	0 026	0 026	G0
SZ Tau	7 1	0 50	0 022	0 023	F8
SU Cas	5 9	0 29	0 014	0 008	F6
RR Lyr	7 2	−0 24	0 223	0 246	A5 s
RR Cet	8 6	−0 26	0 069	0 144	A0
W CVn	10 3	−0 26	0 042	0 192	
SX Aqr	11 8	−0 27	0 027	0 247	
SW Aqr	10 4	−0 34	0 066	0 317	
ST Oph	11 6	−0 35	0 009	0 075	
U Tri	11 6	−0 35	0 023	0 192	
ST Vir	10 8	−0 39	0 018	0 104	
RS Boo	10 3	−0 40	0 014	0 064	A2 s

proper motion, we have the means of the Tables X, VIII and X, IX. Leaving out of consideration cluster-type Cepheids,

[21] H. C. 313, 1927.

we have little indication here of a period-proper motion dependence; and, since the reduced proper motion should be a direct
index of absolute magnitude, the correlation of period and
luminosity is low, as far as this material is concerned.

TABLE X, VIII.—REDUCED PROPER MOTION
FOR INTERVALS OF LOG PERIOD

Mean Log Period	Number of Stars	Mean Reduced P M
		"
1 25	10	0 015*
0 92	11	0 010
0 83	7	0 017
0 71	8	0 014
0 51	6	0 017
−0 32	9	0 176

* W Virginis omitted

TABLE X, IX—LOG P FOR INTERVALS OF
REDUCED PROPER MOTION

Mean Reduced P M	Number of Stars	Mean Log P
"		
0 215	8	−0 24
0 046	8	0 56
0 022	10	0 76
0 013	8	0 97
0 007	7	0 77
0 003	11	0 94

A large systematic observational program on the proper
motions of Cepheid variables is now in progress at the McCormick and Mount Wilson observatories. It may eventually
provide sufficiently accurate material for the direct test of the
existence of a period-luminosity relation for galactic Cepheids,
though observational errors and the peculiar motions of the
stars will make subdivision of the data for this spectral purpose
somewhat precarious. The direct attack is not too hopeful,

but an indirect test is now possible; as shown in the next section, we have a very definite astrophysical way of finding a period-luminosity relation for galactic Cepheids.

51. A Theoretical Period-luminosity Relation for Galactic Cepheids.

—The usefulness of photometric methods, such as the period-luminosity relation, in the measurement of the distances of clusters, star clouds, and nebulae, depends on the uniformity of stellar laws throughout the universe. If Cepheids of the galactic region differ systematically from those in the Magellanic Clouds, the period-luminosity curve may perhaps still be used for clusters and external galaxies, but not safely for isolated Cepheids or galactic star clouds. It is, however, a simple matter, as shown below, to prove definitely that the period-luminosity relation is maintained for galactic Cepheids.

a. From the period-spectrum relation, given above in Section 49, it appears that for Cepheids with periods less than one day the average median spectral class is A6. For stars with periods longer than a day the following values may be read from the period-spectrum curve:

Median Spectrum	Log P	Median Spectrum	Log P
F4	(o 16)	G2	o 79
F6	o 30	G4	1 04
F8	o 43	G6	1.38
Go	o 59	G8	(1 70)

b. In a note on Cepheid variation and the period-luminosity relation,[22] it was pointed out several years ago that if we start with the general gravitational relation $P^2 \propto 1/\rho$, the total luminosity of a Cepheid variable

$$L = \pi r^2 J \tag{1}$$

can be written, with close approximation,

$$L = k(\mu P^2)^{\frac{1}{3}}\Phi(\text{C.I.}) \tag{2}$$

where ρ, P, r, and μ are the mean density, period, radius, and mass, respectively, and the surface brightness J is taken as

[22] Shapley, Mt. W. Contr. 154, 4, 1918.

primarily a function of color index. (Equation (1) is, of course, rigorous; the approximation in (2) arises from writing $J = \Phi(\text{C.I.})$.)

From equation (2) it follows that if the differences in median color for typical Cepheids are ignored—that is, if $J = \Phi$ (C.I.) is a constant—the luminosity will decrease with the product of the mass and the square of the period; and since the range in the masses is believed to be small in comparison with the wide range in period, a common mass may be assumed provisionally, and the absolute magnitude is given approximately by

$$M = a + b \log P$$

with a and b constants. This theoretical relation is of the same form as and accurately represents the observed relation for the Cepheids with periods from 3 to 15 days in the Magellanic Clouds and globular clusters; it amounts to a preliminary theoretical period-luminosity curve applicable to all Cepheids.[23]

It would appear, therefore, that among the galactic Cepheids, where direct observations are difficult because of insufficient data on parallaxes and absolute magnitudes, the intrinsic luminosity should necessarily increase with the periods, as observed in the Magellanic Clouds. But the result obtained in 1917 was admittedly preliminary. Certain critical observations were lacking.

c. The variation in J with P need no longer be neglected, however, and with the observed relation of median spectral class to period, it becomes possible to derive the period-luminosity curve for the galactic Cepheids as a direct consequence of the period-spectrum relation and freed from the assumptions necessary above.[24]

[23] Subsequently, Eddington, Seares, and others have calculated period-luminosity curves on the basis of the pulsation theory of Cepheids; M. N R. A. S., 79, 2, 1918, Mt. W. Contr. 226, 40 ff., 1921. See, also, Shapley, Mt. W. Contr 190, 7, 1920; H. C. 314, 1927.

[24] Russell has independently discussed the problem from another angle and reached much the same conclusions (Mt. W. Contr. 339, 1927), using spectroscopic results by Adams and Joy (Mt. W. Comm. 100, 1927).

A determination such as Eddington has attempted of the actual absolute magnitude of a Cepheid from knowledge of its period alone seems too uncertain for practical applications, mainly because of our insufficient information concerning the masses, the darkening at the limb which affects the mean surface brightness, and the ratio of the specific heats which enters into a more rigorous form of the relation of period to mean density.

The relative intrinsic luminosities, however, can be simply computed for typical Cepheids, since only the ratios of masses, periods, and surface brightnesses are involved, and the uncertain and undetermined factors cancel out or are of the second order. With the relative luminosities determined for a series of galactic Cepheids, the scale of absolute luminosities may be derived from the mean parallax of the nearer variables.

d. Since the surface brightness of a star depends directly on the fourth power of the absolute effective temperature T, which has been determined by several investigators for the various spectral types, we may write, for two Cepheids of different spectral class,

$$\frac{J}{J_0} \propto \frac{T^4}{T_0^4}$$

Then, from equations (1) and (2),

$$\frac{L}{L_0} = \frac{\mu^{\frac{1}{2}} P^{\frac{1}{3}} T^4}{\mu_0^{\frac{1}{2}} P_0^{\frac{1}{3}} T_0^4}$$

and we derive

$$M_0 - M = 10 \log \frac{T}{T_0} + \frac{10}{3} \log \frac{P}{P_0} + \frac{5}{3} \log \frac{\mu}{\mu_0} \qquad (3)$$

where M is the absolute bolometric magnitude. Equation (3) is a definitive relation connecting the absolute magnitude, temperature, period, and mass of a Cepheid variable and involves no assumption beyond Stefan's law and the gravitational relation $P^2 \rho = $ const. Our problem is to solve the equation for M in terms of P.

We could write the first term on the right side of the preceding equation as

$$10 \log \left(\frac{T}{T_0}\right) = 2.5 \log \left(\frac{J}{J_0}\right) = (s_0 - s) \qquad (4)$$

where s is the surface brightness expressed in stellar magnitudes. Various kinds of evidence indicate that the difference in surface brightness from one Harvard class to the next is a little more than one visual magnitude. It will be best, however, not to use an assumed constant value of $(s_0 - s)$ but to deal directly with the temperatures of the Cepheids as indicated by their spectra, since the temperature scale is fairly well known over the range of spectrum here involved.

e. Considering equation (3), we note that, since we are not making use of the data on parallaxes or proper motions (except later to fix a zero point), both M and μ are unknown for any given Cepheid, but P and T can be obtained from observations. The mass-luminosity formula derived by Eddington would give the necessary additional information for a complete solution, but it is possible to keep clear of the theories underlying his formulae and to use instead only the observed mass-luminosity relation in the second approximation, after provisional absolute magnitudes are derived from a preliminary solution.

For a first approximation, therefore, we take the masses of Cepheids as equal, and equation (3) becomes

$$M_0 - M = 10 \log T - 10 \log T_0 + \frac{10}{3} \log P - \frac{10}{3} \log P_0$$
$$= (10 \log T - 37.40) + \left(\frac{10}{3} \log P - 1.97\right) \qquad (5)$$

where T_0 has been set at $5500°$, corresponding to a Class G0 star, and the corresponding value of $\log P_0$ from the period-spectrum curve is 0.59.

f. For the computation (Table X, X) of the relative period-luminosity relation (masses assumed equal) from equation (5), I have used the scale of effective temperatures shown in the first two columns of the table. The adopted values are those deduced from colors, spectrophotometric data, and spectrum analysis on ionization principles. The logarithm of the period is derived from the period-spectrum curve.

TABLE X, X.—COMPUTATION OF A PRELIMINARY PERIOD-LUMINOSITY RELATION

Spectrum	T	Log P	10 Log T − 37 40	10/3 Log P − 1.97	$M_0 − M$
	°				
A0	10000	−0.56	+2.6	−3.8.	−1 2
A5	8500	−0.31	+1.9	−3.0	−1 1
F0	7400	−0.06	+1.3	−2.2	−0 9
F5	6500	+0.23	+0.7	−1 2	−0 5
F7 5	6000	+0 40	+0.36	−0 64	−0 3
G0	5500	+0 59	0 0	0 0	0 0
G2 5	5050	+0 85	−0 4	+0 9	+0 5
G5	4600	+1 22	−0 8	+2 1	+1 3
G7 5	4300	+1 62	−1 1	+3 4	+2 3

A plot of the computed $M_0 − M$, in the sixth column, against the logarithm of the period gives the preliminary period-luminosity curve (bolometric) for galactic Cepheids. The range of three and a half magnitudes in intrinsic bolometric luminosity,

TABLE X, XI.—COMPUTATION OF FINAL THEORETICAL PERIOD-LUMINOSITY
RELATION

Spectrum	M	Log μ	$\frac{5}{3}$Log$\frac{\mu}{\mu_0}$	Revised $M_0 − M$	Revised M	Final M	Log P
A0	−0 7	0 65	−0 4	−1 6	−0 3	−0 3	−0 56:
A5	−0 8	0 67	−0 3	−1 4	−0 5	−0 4	−0 31
F0	−1 0	0 70	−0 3	−1 2	−0 7	−0 6	−0 06
F5	−1 4	0 76	−0 2	−0 7	−1 2	−1 2	+0 23
F7 5	−1 6	0 80	−0 1	−0 4	−1 5	−1 5	+0 40
G0	(−1 9)	0 86	0 0	0 0	(−1 9)	(−1 9)	+0 59
G2 5	−2 4	0 94	+0 1	+0 6	−2 5	−2 6	+0 85
G5	−3 2	1 13	+0 5	+1 8	−3 7	−3 9	+1 22
G7 5	−4 2	1 42	+0 9	+3 2	−5 1	−5 7	+1 62

shown by this computation, indicates that the mass factor in equation (3) is not negligible. In Table X, XI the relevant correction to the magnitude is computed. Corresponding to the provisional absolute magnitude in the second column, derived from Table X, X, the logarithm of the mass is read from the observed mass-luminosity relation as compiled by Eddington.[25] The computed correction, $\frac{5}{3} \log \frac{\mu}{\mu_0}$, gives the revised values of $M_0 - M$ in the fifth column, and the resulting absolute bolometric magnitudes in the sixth column.

These improved values of the absolute magnitudes are used for a new determination of the mass factor and lead to new values of the absolute magnitudes, seventh column of Table X, XI, which are considered final, since a third approximation would not introduce sensible alterations. It is to be noted that the mass-luminosity relation as presented by Eddington is probably subject to modification as more data on masses become available; but a considerable change can be allowed without affecting either the legitimacy of the preceding computation or the final results. Nor will reasonable (and probable) alterations of the assumed zero point of the period-luminosity curve appreciably disturb the trend of the computed relation.

In Table X, XII the final bolometric magnitudes of the seventh column of Table X, XI are reduced to visual and photographic values (the latter with zero point independently adjusted) and compared with observation. The observed visual period-luminosity relation is taken without change from Mount Wilson Contribution 151, and the observed photographic values are from Harvard Circular 280. The former involved mainly the Cepheids in the Magellanic Clouds and in globular clusters; the latter is based altogether on the Cepheids in the Small Magellanic Cloud. The agreement of the values computed by way of the spectrum-period curve of galactic Cepheids with the observed values is surprisingly good. It should be noted incidentally that since the computed values for spectral classes Ao and G7.5 involve extrapolations they are of low weight.

g. The negative residuals $O - C_1$ in Table X, XII for the stars of longer period suggest that the observed visual values

[25] The Internal Constitution of the Stars. 153. 1926.

may be too bright. This is now known to be the case. In the reduction from photographic to visual magnitudes for the original construction of the visual period-luminosity curve,[26] the color correction used for the longer-period Cepheids was too great, as may be verified from the new spectrum-period curve. The visual period-luminosity curve consequently gives the absolute magnitudes too bright for periods greater than ten days; the application of a suitable correction would tend to reduce the corresponding negative residuals $O - C_1$ in Table X, XII.

TABLE X, XII.—COMPARISON OF OBSERVED AND THEORETICAL PERIOD-LUMI-
NOSITY RELATIONS

Spec-trum	Log P	Bolometric Magnitude	Computed Vis M	Observed Vis M	O − C₁	Computed Pg M	Observed Pg M	O − C₁
Ao	−0 56	−0 3.	−0 1.	−0 3	−0 2	−0 3	−0 2.	+0 1
A5	−0 31	−0 4	−0 3	−0 3	0 0	−0 3	−0 3	0 0
Fo	−0 06	−0 6	−0 6	−0 6	0 0	−0 4	−0 ?	0 0
F5	+0 23	−1 2	−1 2	−1 0	+0 2	−0 8	−0 4	−0 1
P7 5	+0 40	−1 5	−1 5	−1 4	+0 1	−1 0	−1 1	−0 1
Go	+0 59	(−1 9)	(−1 8)	−1 8	(0 0)	(−1 2)	−1 4	(−0 2)
G2 5	+0 85	−2 6	−2 4	−2 6	−0 2	−t 7	−1 8	−0 1
G5	+1 22	−3 9	−3 6	−3 9	−0 3	−2 8	−2 6	+0 2
G7 5	+1 62	−5 7	−5 2	−5 4	−0 2	−4 3	−3 5	+0 8

In fact, this necessary correction goes somewhat too far, changing systematically the signs of the residuals. It then appears that a still more satisfactory agreement of both visual and photographic observation with theory would be obtained if we were to introduce Eddington's formulation of the pulsation hypothesis and substitute $P' = P(3\gamma - 4)^{-\frac{1}{2}}$ for P in the foregoing discussion, γ being the ratio of specific heats.

h. In summary, it is found that the period-luminosity curve computed from spectroscopic observations of galactic Cepheids is in close agreement with the observed visual and photographic period-luminosity curves derived from clusters and star clouds. This conclusion is reached independently of data on proper motions, radial velocities, or trigonometric and spectroscopic parallaxes; it is without assumption as to the theory of Cepheid variation. We need have no hesitancy in accepting the comparability of Cepheid phenomena throughout known sidereal

[26] Shapley, Mt. W. Contr 151, 1917

systems or in using the Magellanic period-luminosity curves to measure the distances of galactic Cepheids.

52. The Zero Point.—The foregoing sections of this chapter have shown how definitely we have established the period-luminosity relation for the Magellanic Clouds, for star clusters, and indirectly for the Galaxy. The form of the curve and the deviations from it are now fairly well known, but we remain for the time being in a state of suspense with regard to the zero point. Originally based on the parallactic motions of the few bright Cepheid variables for which accurate proper motions were available, the zero point has been the target of much discussion and suggested revision.

Kapteyn and van Rhijn believed that the large observed proper motions of cluster-type Cepheids in the Galaxy show them to be dwarfs and necessitate an enormous shift of the zero point which I had adopted,[27] with consequent disaster for my measures of galactic dimensions. Later discussions of the proper motions and radial velocities of these variables indicated, however, that they could not be used in the manner adopted by Kapteyn and van Rhijn.[28] Many of the stars of this type belong to the well-known high-velocity group, and when allowance is made for their high space velocities, the proper motion data are not in disagreement with the zero point indicated by the classical Cepheids. Table X, XIII, giving the radial velocities that are now available (mainly from Mount Wilson) for cluster-type variables, shows that the galactic members of this class have extraordinary speed—a circumstance which may be related to anomalies in their variations and is certainly connected with their large proper motions and their wide distribution in galactic latitude. There is nothing in these motions to indicate that the cluster-type Cepheids are dwarfs and the estimated distances of globular clusters seriously wrong.

[27] B. A. N., **1**, 37, 1922.
[28] H. C. 237, 1922; R. E. Wilson, A. J , **35**, 35, 1923.

The most thorough study of the zero point is that by R. E. Wilson.[29] He concluded that the distances that I had based on the period-luminosity relation should be decreased by 20 or 30 per cent; but in view of the now generally accepted Kapteyn

TABLE X, XIII.—RADIAL VELOCITIES OF CLUSTER-TYPE VARIABLES

Designation	Star	Period	Amplitude in Kilometers	Velocity in Kilometers
		d		
001828	SW And	0 44	51	− 16
012700	RR Cet	0.55	85	−102
044515	RX Eri	0.59	40	+ 66
044930	SU Aur	0 47	36	+ 23
045221	U Lep	0 58	49	+114
071531	RR Gem	0 40	51	+ 70
100224	RR Leo	0.45	49	+ 34
113267	SU Dra	0.66	43	−174
121370	SW Dra	0.57	82	− 29
132954	RV UMa	0 47	61	−180
140238	W CVn	0.55	10	+ 20
142932	RS Boo	0 38	75	− 18
162618	VX Her	0 46	31	−390
163358	RW Dra	0 44	30	−109
183932	RZ Lyr	0 51	120	−221
192242	RR Lyr*	0 57		− 60
193056	XZ Cyg*	0 47		−196
205230	UY Cyg	0 56	6	− 36
205515	RV Cap	0 45	70	− 86
211000	SW Aqr	0 46	71	− 20
223564	RZ Cep†	0 31	56	− 14

*Wilson, A. J , **35**, 35, 1923
†Luyten, P. A. S. P., **35**, 69, 1923; see Shapley, H B. 773, 778; Leavitt (and Luyten) H. C. 261, 1924.

correction to Boss' proper motions, this change in the zero point by Wilson may be, as he himself has pointed out, largely effaced. In the most recent discussions of Cepheid motions Oort gets 1.07 ± 0.26 (m.e.) as the correction factor to my parallaxes, using Mount Wilson measures of radial velocity.[30]

[29] A. J., **35**, 35, 1923.
[30] B. A. N., **4**, 91, 1927.

As far as they go, the trigonometric parallaxes in Table X, VI and the Mount Wilson and Upsala spectroscopic parallaxes[31] support the system based on the period-luminosity curve. Van Maanen has found satisfactory agreement of his trigonometric parallaxes[32] with the values from the period-luminosity relation.

In view of Wilson's work, however, and of other indications that the nearer galactic Cepheids give too bright a zero point,[33] I have made a provisional correction of +0.23 magnitudes, which reduces to 0.00 the absolute photographic magnitude of cluster-type variables. The correction has been applied to the period-luminosity curves and to all the computed distances appearing in this volume. It amounts to a systematic decrease of 11 per cent in the distances computed from older period-luminosity curves.

Awaiting the completion in a few years of the McCormick and Mount Wilson investigations of the proper motions of a large number of galactic Cepheids, we adopt this corrected zero point and the scale of distances dependent on it. The resulting absolute magnitudes in clusters and Magellanic Clouds are accordant with a large body of astronomical observations and compatible with recent astrophysical theory. There seems to be no serious inconsistency in the resulting luminosities, except that the revision has made the maximum brightness of stars in globular clusters surprisingly low. I am inclined to predict that the correction to the zero point now adopted will never exceed a quarter of a magnitude and that it may be in either direction; but this prediction should be made with caution because of the increasing evidence of peculiar drifts and of general heterogeneity in the star structure in the immediate vicinity of the sun.

53. The Period-luminosity Relation in Clusters and External Galaxies.

—Figure X, 3 shows clearly that classical Cepheids and cluster-type Cepheids are definitely related by the

---

[31] Shapley, H. C. 237, 1922; Lindblad, Ap. J., 59, 37, 1924.

[32] P. A. S. P., 32, 62, 1920.

[33] See Appendix C for relevant papers by ten Bruggencate, Curtis, Doig, Kienle, Malmquist, and Stromberg.

period-luminosity curve in some of the globular clusters as well as in the Magellanic Clouds. The question has been raised whether the relatively few long-period Cepheids in globular clusters are correctly described as typical. An examination of the observational material leaves no doubt that the stars are normal members of the class; Figure IV, 2 shows that variable star No. 42 in Messier 5 has the usual form of light curve.

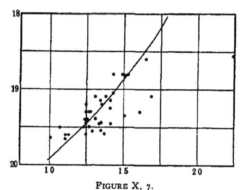

FIGURE X, 7.

Hubble's period-luminosity curve for variables in Messier 31. Coordinates are median apparent photographic magnitudes and logarithms of the periods.

Hubble's work on the Cepheid variables in N. G. C. 6822, a remote star cloud of the Magellanic Cloud type, and in the bright spirals Messier 31 and Messier 33, establishes the period-luminosity relation for those remote external systems.[34]   Figure X, 7 is a period-luminosity curve taken from Hubble's discussion of Messier 31 (the Andromeda Nebula); in his discussions of N. G. C. 6822 and Messier 33 similar plots may be found. The new photographic period-luminosity curve, when applied to these three external systems, gives the following distances:

Messier 31	247 kiloparsecs
Messier 33	236 kiloparsecs
N. G. C. 6822	192 kiloparsecs

[34] Hubble, Mt. W. Contr. 304, 1925; 310, 1926; 376, 1929.

54. Long-period Variables and the Pulsation Hypothesis.—The plot in Figure X, 6 of data given in Table X, V (Section 49) shows a highly significant sequence for the different types of variable stars. The apparently close relation of cluster-type Cepheids and of RV Tauri variables to the classical Cepheids has been generally admitted, but in the period-spectrum relation we appear to link up equally definitely the Cepheid and long-period variables.[35]

In the variability of spectrum and of color, in the form of light curve and its frequent instability, in occasional inconstancy of period, and in the property of high luminosity, the Cepheid and long-period variables are similar. Their principal differences are distribution in the sky and range of variation, but both these differences are apparent rather than real. As to distribution, the selection of stars on the basis of luminosity and distance may be mainly responsible for the differences in galactic concentration. As to range of variation, the radiometric measures of Class M variables by Pettit and Nicholson show that in total radiation the amplitude of the typical long-period variable is not three or four magnitudes, as in visual light, but is of the order of one magnitude, as for cluster-type variables and classical Cepheids.

It appears, therefore, that the pulsation hypothesis, which accounts for cluster-type and classical Cepheid variation more satisfactorily than does any other theory hitherto proposed, may be logically extended to RV Tauri[36] and long-period variation on the basis of the period-spectrum relation and the observed similarities in bolometric absolute magnitude, radiometric variation, spectrum peculiarities, and other properties. Dissimilarities among the four types are only those natural to the different stages of development of the various groups—that is, differences in mean density, in average spectral class, in visual absolute magnitude, and in mean distance and galactic concentration for a given apparent magnitude.

[35] Shapley, H. B. 861, 1928.
[36] Gerasimovič, H. C. 341, 1929.

That the present period-luminosity law does not include the long-period variable stars is admitted; but the law is of an empirical nature. When dependable absolute bolometric magnitudes are available, the red long-period variables may not deviate widely. For class M variables, moreover, with their low densities and temperatures, the internal pulsations would probably be considerably masked; peculiar relations among light, velocity, and spectrum variations should be expected, since the heavily banded spectra may greatly affect photographic, visual, or even radiometric magnitudes. In the mean, the fundamental relation $P^2\rho =$ constant, between period and mean density, appears to hold for long-period variables.

The current investigations of galactic star clouds should eventually show in what manner, if at all, the period-luminosity relation can be extended to long-period variable stars.[37]

[37] Shapley H. Repr 53, 1928; Gerasimovič, H Repr 54, 1928.

CHAPTER XI

THE DISTANCES OF CLUSTERS

To the measurer of the sidereal universe star clusters are beacon lights. They point the way to the center of the Galaxy and to its edges, and throw light on problems of growth and decay. They are parts of the higher galactic organization, deeply involved in its complications and yet so significantly placed that knowledge of their distances is knowledge of the fundamental structure of the Galaxy itself. The globular clusters are a sort of framework—a vague skeleton of the whole Galaxy—the first and still the best indicators of its extent and orientation. The galactic clusters, when near, reveal important drifting tendencies of the stars and, when remote, are frequently the informative nuclei in otherwise indefinite galactic star clouds, thus outlining some of the inner structural detail of the Galaxy. Both types of clusters contribute in fact and in half-hidden intimation to the untangling of the story of sidereal evolution.

Measures of the distances of globular clusters fortunately can be based on Cepheid variables, either directly or through the use of these stars in the calibration of other methods. The galactic clusters, which not only are devoid of Cepheid variables but also lack homogeneity in structure and similarity in absolute dimensions, are more difficult to measure for distance, though much nearer to the earth. In discussing the determination of distances it will be convenient to treat the two types separately.

55. Distances of Globular Clusters Obtained from Cepheids and Bright Stars.—From the period-luminosity curve given in the preceding chapter, distances[1] can be deter-

[1] See special bibliography in Appendix D.

mined directly for all the globular clusters in which Cepheid variables have been studied and magnitude scales determined. The magnitudes of the brighter stars are nearly as useful. In Table XI, I[2] are collected all the relevant observational data now available from my own studies, and also, for N. G. C. 5053 and N. G. C. 5466, the magnitudes measured by Baade with the Bergedorf reflector. For the sake of homogeneity, the magnitudes measured at Bonn by Küstner for three globular clusters (M 3, M 15, and M 56) were not used.

a. The Observations.—The new material represents a considerable expansion over that in hand 12 years ago. The determination, in 1917, of the parallaxes of 68 globular clusters included only 7 for which the variable stars had been studied, and for two of these the preliminary data could not be used quantitatively. As the variable stars in clusters are fundamental in calibrating methods of determining distances, I have given considerable attention since 1917 to the discovery and observation of variable stars in clusters. Mount Wilson plates and various series in the Harvard collection have been used for this work. I am indebted to several assistants at Mount Wilson and Harvard for aiding in this laborious research and especially to Miss Sawyer, who has taken an active part in the recent revision of cluster distances.[3]

There are now 19 instead of 5 clusters in which variable stars have been measured sufficiently to enter the new determination of distances. Of the 730 variable stars in these 19 clusters, 524 have been studied enough to be useful in fixing the "median magnitudes" for the clusters concerned.

In 1917 the absolute magnitudes of the high-luminosity stars had been measured for only 28 clusters; we now have measures on the brighter stars in 48 systems.

b. The Methods.—For a new determination of distance and distribution, I have followed in principle, though not in detail, the methods developed and described in 1917. As before, the

[2] From H. B. 869, 1929.
[3] H. B 869, 1929.

reasonable assumption is made that the median absolute magnitude of variables with periods less than a day is constant from cluster to cluster. The difference between the median magnitudes of the variables and the magnitudes of the brightest stars is again found, on the average, to be so definite a quantity that brighter star magnitudes themselves can be used as criteria for those clusters where variable stars are lacking or have not been analyzed. We go farther by taking measures of integrated apparent brightness (total magnitude) and angular diameter as criteria of relative distance, using these measures, after appropriate calibration, not only to strengthen the determination for the 48 clusters whose variables and bright stars have been studied in some detail (Table XI, I) but also for the other 45 clusters now known in the galactic system for which we have as yet no measures of variables or of individual bright stars.

c. *The Results.*—A few of the essential details of Table XI, I should be mentioned. The classes in the second column are those described in an earlier chapter and listed for all globular clusters in Appendix A. The apparent median magnitude of cluster-type variables is in the third column. When a cluster contains long-period Cepheids, their magnitudes are reduced to the cluster-type median by means of the period-luminosity relation and are combined with the magnitudes of the cluster-type variables.

Since we have adopted zero as the absolute photographic magnitude for cluster-type Cepheids, the third column actually contains a direct determination for 19 clusters of the distance modulus $m - M = 5$ $(\log d - 1)$, where d is the distance in parsecs and m is the apparent median photographic magnitude.

The parenthetical numbers in the third column are the combining weights assigned to the mean median magnitudes of the variables in each cluster. These weights depend on the number of variables, on the detail with which periods and light curves are known, and on the estimated accuracy of the magnitudes.

TABLE XI, I.—MAGNITUDES OF VARIABLES AND BRIGHT STARS

N. G. C.	Class	Photographic Magnitude				Preliminary Modulus	Notes
		Variables	25 Br	6th Star	30th Star		
104	III		13 09	12 4	13 4	14 33	Note 1, 47 Tucanae
288	X		14 80	14 5	15 1	15 87	Note 2
362	III	15 5 (4)	14 12	13 5	14 8	15 49	
1904	V		15 29	15 01	15 72	16 61	Messier 79
2808	I		14 9	14 3	15 4	16 25	
3201	X	14 52 (5)	13 52	13 3	13 8	14 57	Note 3
4147	IX	16 8 (1)	16 58	16 23	16 93	16 99	GP
4590	X	15 90 (6)	14 80	14 31	15 08	15 87	Messier 68
5024	V		15 07	14 94	15 26	16 36	Messier 53
5053	XI	16 19 (8)	15 65	15 1	16 0	16 15	GP
5139	VIII	14 37 (5)	12 91	12 6	13 1	14 22	ω Centauri
5272	VI	15 50 (8)	14 23	13 92	14 45	15 48	Messier 3
5466	XII	16 17 (6)	15 72	15 1	16 2	16 16	GP
5897	XI		15 15	14 9	15 4	16 16	
5904	V	15 26 (8)	13 97	13 74	14 27	15 26	Messier 5
6093	II		14 88	14 72	15 09	16 24	
6121	IX	14 27 (6)	13 88	13 3	14 4	14 29	Note 4, GP, Messier 4
6144	XI		15 76	15 2	16 3	16 22	GP
6171	X		15 46	15 2	15 9	16 57	
6205	V	15 20 (4)	13 75	13 45	13 92	15 10	Messier 13, Note 5
6218	IX		13 97	13 56	14 31	15 07	Messier 12
6229	VII		16 18	15 90	16 37	17 36	
6235	X		16 17	15 7	16 8	17 28	
6254	VII		14 06	13 35	14 38	15 17	Messier 10
6266	IV	16 40 (6)	15 87	15 6	16 1	16 37	Irreg, Messier 62
6333	VIII		15 61	15 08	15 88	16 70	Messier 9
6341	IV		13 86	13 60	14 16	15 18	Messier 92
6356	II		17 16	16 86	17 44	18 51	
6397	IX		12 61	11 9	13 1	13 67	GP
6402	VIII		15 44	14 85	15 86	16 56	Messier 14
6535	XI		15 9	15 3	16 4	16 89	
6541	III	14 42 (4)	13 35	12 7	13 8	14 53	Note 3
6626	IV		14 87	14 49	15 11	16 14	Messier 28
6638	VI		16 22	15 90	16 60	17 48	
6656	VII	14 06 (5)	12 93	12 80	13 26	14 12	Messier 22
6712	IX		16 10	15 65	16 36	17 17	
6723	VII	15 33 (6)	14 20	13 7	14 8	15 37	
6752	VI		13 26	12 8	13 6	14 47	
6779	X		15 31	14 98	15 70	16 39	Messier 56
6809	XI		13 58	12 9	14 2	14 58	Messier 55
6864	I		17 06	16 76	17 35	18 43	Messier 75
6934	VIII		15 78	15 33	16 11	16 91	
6981	IX	16 80 (8)	15 86	15 53	16 20	16 86	Messier 72
7006	I	18 96 (6)	17 50	16 99	17 89	18 91	
7078	IV	15 63 (7)	14 31	14 13	14 55	15 63	Messier 15
7089	II	15 71 (4)	14 61	14 25	14 76	15 81	Messier 2
7099	V		14 63	13 77	15 04	15 80	Messier 30
7492	XII	.	16 82	16 3	17 1	17 22	GP

NOTES TO TABLE XI, I

1 The long-period variables in 47 Tucanae (H. B. 783) could not be safely used in measuring the distance.

2 The mean magnitude of the 25 brightest stars was determined at Mount Wilson to be 14 81, with a range of 14 38 to 15 04.

3 The magnitudes of N G C 3201 and N. G. C 6541 may be considerably in error due to unsatisfactory comparison sequences. They are not included in the determination of the reduction curves for apparent integrated magnitudes and diameters, though they are appropriately used in constructing Table XI, II

4. For N. G. C. 6121 the zero point of the magnitude scale depends on both Harvard and Mount Wilson measures of Mount Wilson plates.

5. The mean magnitude of the 25 brightest stars was determined at Harvard to be 13.76 with a range of 13 4 to 14 1.

The means of the third column are used in conjunction with the distance moduli derivable from the next three columns to determine the preliminary modulus in the seventh column.

The fourth column of Table XI, I contains the mean magnitude of the 25 brightest stars (after the exclusion of 5 of maximum brightness in order to avoid, or at least to diminish, the effect of optical doubles and of the chance superposition of bright field stars). A weakness in using this method lies in the dependence of the area in each cluster surveyed on the compactness or distance of the cluster, or on the richness of the foreground of galactic stars. Uniformity of selection was strenuously sought, and we are convinced that for most clusters the mean of the 25 brightest stars, as well as the magnitude of the thirtieth star, would not be appreciably disturbed by including or excluding too much of the dense central region.

In the fifth and sixth columns are given the apparent magnitudes of the sixth and thirtieth stars in each cluster; the sixth star is the brightest object and the thirtieth the faintest included in the means of the 25 brightest.

The difference between the median magnitudes of the cluster-type variables (third column) and the magnitudes of the three succeeding columns are found to depend on the class of the cluster. Table XI, II gives the readings from the smooth curves that represent the change of the difference with class. In the earlier work, a single constant value, 1.28, for the difference between the median and the 25 brightest, was used, and the two other sets of differences were not considered in deriving distances. The range now found in med.−25 br. is from 0.92 (Class XII) to 1.34 (Class I).

The present method of using three different points in the sequence of apparent magnitudes is essentially equivalent to comparing, from one cluster to another, the general luminosity curves for giant stars. Its advantage over previous practice lies in the allowance it makes for abnormal distribution—or even small deviations from the average—in limited groups of stars. The method is obviously justified by the non-parallelism of the three curves representing the relation of reduction factors to class of cluster

A comparison of the modulus from variable stars alone (third column of Table XI, I) with the preliminary modulus (seventh column) based on bright stars and variables together, indicates how satisfactorily the data for bright stars agree with the results from the variables. The agreement shows, in fact, to what extent one typical cluster is comparable with another inside the various classes.

TABLE XI, II.—REDUCTION TO MEDIAN MAGNITUDE OF CLUSTER-TYPE VARIABLE STARS

Class of Cluster	Median — 25 Br	Median — 6th Star	Median — 30th Star
I	1 34	1 77	1 04
II	1 33	1 74	1 01
III	1 32	1 71	0 98
IV	1 30	1 68	0 95
V	1 28	1 64	0 92
VI	1 24	1 60	0 89
VII	1 20	1 56	0 86
VIII	1 15	1 51	0 83
IX	1 10	1 46	0 80
X	1 05	1 40	0 77
XI	0 99	1 34	0 74:
XII	0 92.	1 29	0 71.

d. Giant-poor Clusters.—A number of the clusters of the more open classes (classes IX to XII) were found upon examinations for star frequency to be poor in giant stars. Their luminosity curves are abnormal. Such objects are indicated by the letters GP in the last column of Table XI, I. They were not used in deriving differences in Table XI, II for normal clusters, and the irregular cluster Messier 62 (N.G.C. 6266) was also excluded

The GP clusters with measured variables were used, however, to determine special reduction factors for all the "giant-poor" clusters. A study of these abnormal systems shows that the following mean values can be satisfactorily used in reducing the magnitude measures to the standard median magnitude of the cluster-type variables:

Median — Mean of 25	=	+0 44
Median — Sixth brightest	=	+0 94
Median — Thirtieth brightest	=	+0 03

e. The System of Weighting.—For the typical clusters for which bright star magnitudes have been measured (Table XI, I), the reductions to "median magnitude" are made directly with the aid of Table XI, II, and the three resulting determinations for each cluster are combined with weights 2, 1, 1, for the mean of 25, sixth, and thirtieth, respectively, to get the preliminary distance modulus. The modulus from variable stars, when available, was, of course, included with appropriate weight in the mean value in the seventh column of Table XI, I. The final weight of each preliminary modulus is, therefore, the weight in the third column (from the variable stars) increased by four.

56. Distances of Globular Clusters Obtained from Diameters and Integrated Magnitudes.—Further steps in deriving the distances of the clusters are now obvious and need only be summarized. Plotting the values of the preliminary modulus in Table XI, I against the integrated brightness and angular diameter,[4] we get two empirical curves that may be used for the derivation of the distance modulus of any globular cluster for which the total magnitude and angular dimensions have been measured.

a. Angular Diameters.—The adopted diameters, which are given for all globular clusters in Appendix A, are based on measures made at Harvard on photographs of Series A (Bruce 24-inch refractor) and of Series AX and AY, which are made with short-focus cameras; double weight is assigned to the large-scale plates. It should be noted that the measured angular diameter of a cluster depends on exposure time; the recorded diameters *d* are, therefore, not exactly proportional to the parallaxes nor

[4] H. B. 852, 1927.

related by the normal formula $d = k10^{0.2m}$ to the integrated apparent magnitudes m. This limitation, however, does not decrease their value as a measure of relative distances.

In general, the measures of diameter refer to the nucleus or the main body of the cluster. Plates of long exposure, made with large telescopes, when carefully counted and analyzed, show that the clusters are of considerably greater extent than is recorded in these "surface" measures of angular diameter. The distribution of cluster-type variables also frequently indicates the wide dispersion of cluster stars. For example, Bailey notes the existence of variables 19' from the center of N. G. C. 3201,[5] though the "working diameter" in the catalogue is 7'.7, in agreement with the value given in the New General Catalogue.

b. Integrated Apparent Magnitudes.—The photographic magnitudes for globular clusters are on a convenient but not the conventional scale. They were measured by Miss Sawyer on plates of the AX and AY series.[6] The scale is much more open than in the customary Pogson system, and as a result the integrated brightness listed in Appendix A ranges from the third to nearly the thirteenth magnitude. If the scale were on the usual system of stellar magnitudes, this difference of 10 magnitudes would indicate a factor of 100 in the relative distances, rather than the factor of 10 which is actually found.

The measures of brightness are, however, fairly accurate, since much care was taken in the selection of plates and of magnitude sequences. Photographic images of globular clusters depend on the lenses, plates, and photographic development, and for the brighter and larger clusters are necessarily uncertain not only because of the size and texture of the photographic image but also because of the scarcity of suitable comparison stars and the general weakness of photographic sequences for bright magnitudes. With the warning that these new measures of brightness cannot be used for the computation of absolute

[5] H. C. 234, 1922.
[6] H. B. 848, 1927.

magnitudes or for comparison with visual integrated magnitudes for the determinations of color index or, as Dufay[7] has discovered, used directly for the derivation of relative distances, we can, however, proceed to make important indirect use of them, when properly calibrated, to get at the distances of clusters for which we have no data on individual bright stars or variables.

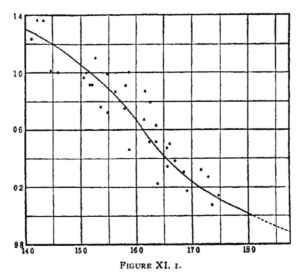

FIGURE XI, 1.

Relation of logarithm of diameter to distance modulus (abscissae) for globular clusters

c. The Distance Moduli and Their Weights.—The two curves which were used for the derivation of distance moduli from measures of diameters and of total magnitudes are shown in Figures XI, 1 and XI, 2. Their coordinates are contained in Tables XI, III and XI, IV. The actual moduli derived for all 93 clusters are omitted[8] from Appendix A, but they may be recovered if desired from the curves or tables of this chapter and from the tabulated measures in the appendix.

[7] Bul. Obs. Lyon, 11, 59, 1929.
[8] They are tabulated in Table III, p. 7, of H. B. 869, 1929.

Unit weight is assigned to each of these new determinations of cluster distances. When other values are available, as in 48 of the clusters, the weight 4 is assigned to the modulus depending on the bright stars, and the weight for the modulus from variable stars is that given in Table XI, I above. For instance, the four values of the modulus for Messier 3 (N. G. C. 5272) are 15.50 (variable stars), 15.45 (bright stars), 15.23 (diameters), and 15.0 (magnitudes). The corresponding weights are 8, 4, 1, 1. The adopted mean modulus is 15.43, corresponding to a mean distance of 12.2 kiloparsecs or about 40,000 light years.

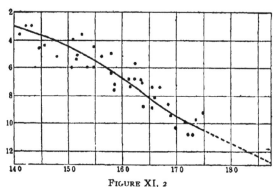

FIGURE XI, 2

Relation of integrated magnitude to distance modulus (abscissae) for globular clusters.

For the "giant-poor" clusters the modulus from the angular diameter is derived from the curve for normal systems (Figure XI, 1) and the modulus from total magnitude is derived from a smoothed curve based on the integrated magnitudes for only those clusters that are deficient in giant stars. Loose globular clusters, such as N. G. C. 288 and N. G. C. 3201, may approach the deficient condition, but the variable reduction factors of Table XI, II largely correct for minor systematic deviations from average conditions of magnitude frequency.

The adopted mean modulus for each cluster is given in Appendix A, followed by a letter indicating the quality of the determination. The assumed quality depends on the final weight of the modulus and the accordance of the various determinations. The letter "a" indicates the values of highest weight; the letter "e" refers to the most uncertain determinations, which are, unfortunately, still too numerous. The distribution among the qualities is: a 13, b 25, c 23, d 17, and e 15.

TABLE XI, III —MODULUS-DIAMETER CURVE

Modulus	Log Diameter
14 0	1 30
14 5	1 20
15 0	1 05
15 5	0 89
16 0	0 67
16 5	0 42
17 0	0 24
17 5	0 11
18 0	0 02
18 5	9 92

TABLE XI, IV.—MODULUS-MAGNITUDE CURVE

Modulus	Integrated Magnitude
14 0	3 1
14 5	3 7
15 0	4 5
15 5	5 6
16 0	6 7
16 5	8 1
17 0	9 4
17 5	10 5
18 0	11 4·
18 5	12 2.

The seven clusters for which the N. G. C. numbers are marked with daggers in the first column of the catalogue (two of them also appear in Table XI, I) may possibly be galactic clusters or nebulous groups rather than normal globular clusters. They are described in Chapter II, Section 6. An exponent n in the first column refers to the notes at the end of the catalogue.

For clusters where there is good material on variable stars, the determinations based on total magnitude and diameter contribute but slightly to the finally adopted modulus. The computed distances for nearly half of the clusters, however, depend entirely on the relatively low weight determinations of the apparent brightness and diameter. Efforts will be made within the next few years to extend the work on variables and magnitudes of individual stars. As a result, alterations in the distances of individual clusters can confidently be expected, but it is practically certain that the scale of the system of clusters and of the Galaxy will not be affected thereby. The zero point correction predicted in the last chapter is the only agent likely to disturb the general scale of distances.

d. Comparison with Earlier Results.—As previously noted, we have made a systematic correction of 11 per cent to the distances of globular clusters through making a provisional alteration in the zero point of the period-luminosity curve. On comparing the distances now given (Appendix A) with those previously obtained through my investigations at Mount Wilson,[9] the average difference is found to be 12 per cent (after allowing for the systematic change). The revision of the individual distances has therefore not been at all drastic, though in a few cases where the early material was unexpectedly weak it has been more than 30 per cent. Because of the great increase in the basic photometric data and the number of globular clusters with Cepheid variable stars involved, the present values are much more secure than those formerly determined.

At times during the past 12 years the scale of distances has been challenged, and evidence or argument advanced to show that I had derived cluster distances and galactic dimensions that might be from five to one hundred times too large. It is inadvisable to take space to reproduce here or even to summarize these many discussions, because the general order

⁹ Mt. W. Contr. 152, 1918.

of distances, and consequent galactic arrangement, is now very generally accepted. It should suffice to mention ten Bruggencate, Charlier, Crommelin, Curtis, Doig, Hopmann, Kapteyn and van Rhijn, Lundmark, van Maanen, Malmquist, Oort, Parvulesco, Perrine, Schouten, Seares, and R. E. Wilson as principal contributors on one side or the other of the discussion and for the details refer to their papers in the general bibliography (Appendix C).

57. A Working Catalogue of Galactic Clusters (Appendix B).—The discussion of the number and distribution of galactic clusters in Chapter II, Section 7, was based on a new and fairly homogeneous catalogue which is given in Appendix B. Although intentionally incomplete, because of the adopted restrictions which exclude poor or indefinite groups, and perhaps overlooking a few clusters that fall within these limitations, the catalogue is probably the most serviceable yet compiled for the general study of galactic clusters. Miss Payne is responsible for the classification of the individual clusters and for the estimates of magnitudes and of numbers of stars. The classifications (sixth column of the catalogue) are on the system proposed in Chapter II. The galactic coordinates are on the Harvard system (Pole at 12^h40^m, $+ 28°$).[10] The angular diameter is from Melotte's catalogue, except when italicized, in which case the estimate was made by Miss Roper using Harvard photographs; it is necessarily approximate and generally refers to the obvious nucleus. Detailed star counts nearly always extend the diameters.

The orientation, expressed as the position angle of the major axis of an elongated cluster with reference to the galactic equator, was estimated independently by Miss Payne and the writer on Harvard photographs for the more compact galactic clusters (Melotte's Class II); the results, which are very accordant, are discussed in Chapter VII. The ninth column of the catalogue indicates the approximate number of stars that could be

[10] Pickering, H. A., **56**, 1, 1912.

assigned to each cluster on plates whose fainter magnitude limits are as given in the following column. In the clusters for which this fainter limit is not given, the majority of the cluster stars are much brighter than the limit.

It is probable that many stars within the bounds of the cluster are superposed members of the intermediate galactic field, for these systems usually lie in rich regions in low galactic latitude. Obvious bright foreground stars were not considered in choosing the fifth star for each group and estimating its magnitude. It is probable, therefore, that nearly always the estimated magnitude in the eleventh column actually refers to a star that is near the maximum luminosity in its cluster. How bright absolutely that object may be depends, to some extent, on whether it is a member of a Pleiades or a Hyades type of cluster, and also on whether the cluster is poor or rich in stars. No claim to accuracy is made for these estimated magnitudes; they are given as rough indicators of the brighter limit of apparent magnitude and as a means of making preliminary estimates of the distances and space distribution of galactic clusters.

58. Parallaxes of Galactic Clusters.—Direct trigonometric measures must necessarily fail to give useful information on the distances of galactic clusters, even when they are as near as the Pleiades. Measures of proper motions, and fairly extensive studies of the spectral composition of some of the nearer galactic clusters, have led to useful estimates of the approximate distances. For ten systems, including the Pleiades, the Hyades, Praesepe, Messier 11, Messier 37, the double cluster in Perseus, and the bright cluster in Coma Berenices, the distance in kiloparsecs has been determined through more or less detailed studies of motions, magnitudes, and spectra, and is entered between the twelfth and thirteenth columns of Appendix B. The sources are given in the notes at the end of the catalogue. The accuracy of these ten values is not high, except for the Hyades.

For other galactic clusters no equally dependable measures of the distances are yet available, though provisional photometric or spectral parallaxes have been published by various investigators. Doig[11] and Raab,[12] in particular, have analyzed the spectral data and derived useful preliminary estimates for many of the brighter groups. My own values for a number of the galactic clusters[13] are systematically too great; the published estimates were admittedly very provisional and gave distances that now appear on the average to be two or three times too large because of the tentative assumption that the brighter stars were of exceptionally high luminosity.

Trumpler's spectroscopic and photometric researches on galactic clusters, which have been in progress for some years, should eventually give fairly accurate values of the distances of many of the galactic clusters; his method involves the use of luminosity curves for various spectral classes in the clusters[14] or, what is essentially the same, the use of a Russell diagram for fixing the distance modulus. The final standardization of his system of distances will probably await much serious work on the absolute luminosity dispersion for stars of Class A.

Spectroscopic parallaxes should eventually give the distances of a number of galactic clusters which contain late type stars; and with the development of dependable spectroscopic methods for early type stars, such as those foreshadowed by Miss Williams[15] in analyses of absorption lines in Class A spectra, the spectrum-line method may turn out to be the most dependable one for measuring the distances of galactic groups. It will be a procedure much less time-consuming than the Russell diagram method.

Since the accurate determination of the distances of galactic clusters is still mainly in the future, it seems worthwhile for the time being to tabulate direct photometric estimates.

[11] J. B. A. A., **35**, 201, 1925.
[12] Lund Medd. Ser. 2, 28, 1922.
[13] Mt. W. Comm. 62, 1919.
[14] P. A. S. P., **37**, 307, 1925.
[15] H. C. 348, 1929.

Assuming that the fifth star in order of brightness in a galactic cluster has an absolute magnitude like that of an average bright Class A star, +0.5, we derive the distances given in the twelfth column of Appendix B.   In the thirteenth column are given the distances corresponding to an assumed absolute magnitude of −0.5.   Since we are dealing with objects selected on account of high luminosity, these greater distances are probably more nearly correct.[16]   They have, therefore, been used in computing, in the next two columns, the linear diameter of each cluster in parsecs and the distance of the cluster from the adopted galactic plane.   The wide range in linear diameters, reflecting, in part, the difficulty of estimating the bounds of a galactic cluster, is a striking feature of these computed results.   The smallest objects appear to be only a few light years in diameter, and the largest more than 50.   Not much weight can be put on individual values of the distance, but it is practically certain that these values are of the right order of magnitude and can serve to give a correct idea of the distribution of galactic clusters in space.   The light they throw on galactic dimensions is considered in the following chapter.

[16] The assumption that the fifth star in the cluster is not more luminous than −0.5 implies, in general, that this star is not earlier in spectral class than B8. For a cluster of the spectral constitution and richness of the Pleiades, this assumption would, therefore, give too small a distance. Trumpler has found that about half the clusters he has classified are of the Pleiades type, but, as pointed out in Section 15, this proportion is probably too large because of observational selection. The clusters of the Pleiades type are apparently poor in luminous stars of early type, for of fifteen enumerated by Raab, not one has a *fifth* star of class as early as B5. The adopted method, therefore, probably does not lead to serious systematic error.

CHAPTER XII

DIMENSIONS OF THE GALAXY

The radius of the space-time world, the total mass of the universe, the comparability of galaxies—these basic problems involve directly the measured dimensions of our galactic system. The appeal of such deep and generally unanswerable questions has encouraged the attempt to find the extent of the Galaxy from measures of its star clusters. The attempt succeeds in the gross but not in detail. The clusters dimly outline the size but show little of the structure. We may confidently expect that analyses of star motions, star clouds, and individual stellar distances will in time reveal with more than present clarity the significance of our galactic system in the total material universe.

59. Membership in the Galaxy.—By the term "Galaxy," or "the galactic system," is meant the aggregate of stars and nebulae for which the distributions appear to be organized with respect to the galactic plane. Globular clusters are therefore included, with galactic stars and galactic clusters and nebulae; but the Magellanic Clouds and the extra-galactic nebulae (spiral nebula family) are outside the organization.

Possibly, however, some of the remote globular clusters (e.g., N. G. C. 7006, N. G. C. 4147, and Messier 75) are actually independent, being either fugitives from the Galaxy or chancing for the moment (cosmically speaking) to be moving in this part of space. Some of the high-velocity stars of the sun's neighborhood also may eventually escape from galactic control. We need more information on speeds and masses and on the phenomena of galactic rotation before we can pass judgment on these questions of membership.

The complete freedom of the Magellanic Clouds from our Galaxy is but a surmise based on relative masses, present positions, and radial velocities. Without accurate information concerning their proper motions, we cannot safely assume from radial components that they are receding from the Galaxy[1]; in fact, the increasing evidence that the Galaxy is in rapid rotation argues for the affiliation of the Magellanic Clouds with our Galaxy, or, at least, with a local group of galaxies that would also include the three Andromeda nebulae, Messier 33, and some others.[2]

In measuring the Galaxy, we should at the start admit indefinite limits, and also striking irregularities, not only in the interior but probably also at the edges. The dimensions discussed below are therefore not to be taken too literally as marking the boundaries or even as giving sharp limits of star density. At best, we measure or estimate the distances of the remotest attainable stars or groups of stars which yield to present methods and which appear to be members of the Galaxy.

60. The Higher System of Globular Clusters.—To illustrate the space distribution of the 93 known globular clusters of the galactic system (Appendix A), the following rectangular coordinates have been computed for all clusters:

$$X = R \cos (\lambda - 327°) \cos \beta$$
$$Y = R \sin (\lambda - 327°) \cos \beta$$
$$Z = R \sin \beta$$

where R is the distance in kiloparsecs, β is the galactic latitude, and $(\lambda - 327°)$ is the galactic longitude measured from the direction to the center of the cluster system. The latitude, longitude, distance, and $R \sin \beta$ are given in Appendix A; the computed quantities X and Y are given in Table XII, I.

a. Eccentric Position of the Solar System.—A diagram of the distribution of the globular clusters in the plane of the Galaxy

---

[1] Luyten, H. C. 326, 327, 1928.
[2] Shapley, H. Repr. 61, 1929.

(XY plane) is shown in Figure XII, 1. Crosses indicate clusters lying on the north of the galactic plane, and dots those on the south; the smaller the symbol, the more distant is the object from the plane. The equality of the division by the galactic plane of the supersystem of clusters is remarkable—47 clusters are on the north, 46 on the south.

TABLE XII, I.—COORDINATES OF GLOBULAR CLUSTERS

N G C.	$R \cos \beta \cos (\lambda - 327°)$	$R \cos \beta \sin (\lambda - 327°)$	N G C	$R \cos \beta \cos (\lambda - 327°)$	$R \cos \beta \sin (\lambda - 327°)$	N G C.	$R \cos \beta \cos (\lambda - 327°)$	$R \cos \beta \sin (\lambda - 327°)$
104	+ 2 8	− 3 9	6139	+27 7	− 8 5	6517	+47 0	+17 1
288	− 0 5	− 0 1	6144	+17 3	− 2 4	6522	+35 8	+ 0 6
362	+ 4 5	− 7 5	6171	+19 4	+ 1 7	6528	+44 0	+ 1.2
1261	0	−13 7	6205	+ 4 0	+ 6 8	6535	+23 0	+12 3
1851	− 5 0	−10 6	6218	+ 9 5	+ 2 9	6539	+35 8	+13 8
1904	−12 3	−13 2	6229	+ 6 7	+21 8	6541	+ 8 6	− 1 5
2298	−10 4	−23 4	6235	+28 0	0	6553	+26 6	+ 2 6
2419	−28 0	− 0 5	6254	+10 0	+ 2 9	6569	+29 1	+ 0 5
2808	+ 3 0	−15 6	6266	+18 3	− 1 6	6584	+18 6	− 5 9
3201	+ 1 1	− 9 0	6273	+16 0	− 0 7	6624	+21 7	+ 1 2
4147	− 0 9	− 4 5	6284	+27 6	− 0 5	6626	+16 2	+ 2 3
4372	+ 5 0	− 8 0	6287	+27 4	+ 0 5	6637	+18 4	+ 0 6
4590	+ 6 7	−10 7	6293	+22 8	− 0 8	6638	+29 0	+ 4 3
4833	+ 8 8	−13 0	6304	+25 0	− 1 8	6652	+23 0	+ 0 6
5024	+ 3 2	− 1.3	6316	+31 5	− 0 8	6656	+ 6 3	+ 1 2
5053	+ 3 5	− 1 0	6325	+46 0	+ 0 4	6681	+18 7	+ 1 0
5139	+ 4 2	− 5 0	6333	+20 3	+ 2 1	6712	+23 2	+11 6
5272	+ 2 0	+ 1 7	6341	+ 3 3	+ 8 6	6715	+18 6	+ 1 9
5286	+16 0	−17 0	6342	+39 1	+ 4 1	6723	+11 7	+ 0 2
5466	+ 3 8	+ 3 4	6352	+18 4	− 6 0	6752	+ 6 7	− 3 0
5634	+19 8	− 5 8	6356	+48 5	+ 6 0	6760	+23 1	+16 8
I4499	+14 0	−18 0	6362	+12 0	− 8 0	6770	+ 9 1	+17 8
5824	+24 4	−11 9	6366	+26 3	+ 9 1	6809	+ 7 9	+ 1 2
5897	+15 0	− 4 0	6388	+16 6	− 4 2	6864	+40 7	+14 8
5904	+ 7 4	+ 0 6	6397	+ 5 1	− 2 1	6934	+14 1	+18 7
5927	+16 8	−11 0	6402	+17 7	+ 7 2	6981	+15 6	+11 4
5946	+27 2	−17 0	6426	+31 3	+17 4	7006	+22 4	+48 0
5986	+15 4	− 6 2	6440	+49 5	+ 7 0	7078	+ 4 7	+10.6
6093	+16 5	− 1 9	6441	+21 0	− 0 2	7089	+ 6 6	+ 9.2
6101	+14 7	−13 5	6453	+50 0	− 0 4	7099	+ 8 6	+ 4 6
6121	+ 6 9	− 1 0	6496	+21 2	− 4 5	7492	+ 6 4	+ 9 1

The direction to the center of the system, derived in Chapter II from the apparent positions of globular clusters, is seen to agree with the direction on the basis of space coordinates; in Figure XII, 1 there are 46 positive values of Y and 46 negative values, with $Y = 0$ for one cluster. (The remote system

N. G. C. 7006, with coordinates $X = 22.4$, $Y = 48.0$, falls outside the limits of the diagram.)

The origin of coordinates for Figure XII, 1—that is, the position of the observer—is on the border of the globular cluster system. The center of gravity of the system of clusters, indicated by an open square, has the coordinates $X = +16.4$, $Y = -0.3$ (or $Y = +0.2$ if the remote and isolated N. G. C. 7006 is included).

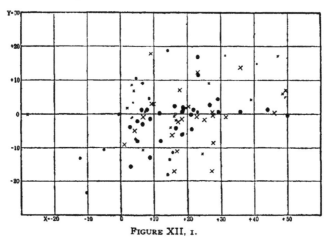

FIGURE XII, 1.

Distribution of globular clusters in the plane of the Galaxy. The sun is at the origin of coordinates.

Probably 10 or 20 globular clusters, within the limits of space represented by the diagram, await discovery. Obscuring nebulosity possibly conceals most of these systems, of which the existence is intimated by the scarcity of observed points in the right half of the figure. Of course, we need not assume high regularity in distribution, or even approximate circularity in the projected array; but it is probably observational difficulties caused by nebulosity and by the small dimensions and faint magnitudes of remote clusters, and not inherent irregularities, that have produced the apparent incompleteness for X greater than $+30$ kiloparsecs. On the left side of the diagram, from

$X = 0$ to $X = -20$, where the survey may be considered
sufficiently exhaustive, the complete absence of clusters from
one quadrant is even more striking and structurally significant
than the scarcity of values of X greater than $+30$.

The cluster at the extreme left is N. G. C. 2419—an object
in a region far from other clusters, found through studies of the
Lowell Observatory photographs.[3]

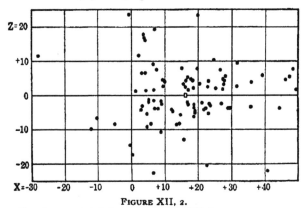

FIGURE XII, 2.
Distribution of globular clusters in the XZ plane.

b. The "Region of Avoidance."—The same asymmetrical
position of the sun with respect to the supersystem of globular
clusters is shown in Figure XII, 2, where all 93 systems are
plotted on the XZ plane. The center of gravity—that is,
the algebraic mean values of X and Z, indicated by an open
square—is at $X = +16.4$, $Z = +0.4$. N. G. C. 2419 again
stands out on the extreme left.

The most interesting feature of Figure XII, 2, which represents a section perpendicular to the galactic plane, is the "region
of avoidance." The scarcity of globular clusters in low galactic
latitudes, shown most clearly in Figure 4 of Chapter II, is
again in evidence. Here is a central section 2.5 kiloparsecs
(8,000 light years) in diameter, in which no globular cluster
has been found. On the other hand, there is only one galactic

[3] Shapley, H. B. 776, 1922.

cluster out of the 249 listed in Appendix B that does not fall well within this mid-galactic segment, and that cluster, N. G. C. 2243, is of doubtful nature and uncertain distance. Practically all known galactic stars and nebulae also fall within this "region of avoidance."

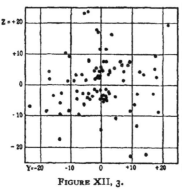

FIGURE XII, 3.
Distribution of globular clusters in the YZ plane.

c. Projection on the YZ Plane. Figure XII, 3 shows the distribution of globular clusters on the YZ plane, which is perpendicular to the line joining the sun and the center of the system at galactic longitude 327°, galactic latitude 0°. The "region of avoidance" is again clearly shown, and also the essential symmetry of the globular cluster system, for the numbers of clusters in the four quadrants are 23, 24, 22, and 24. Again N. G. C. 7006 is outside the diagram, with coordinates $Y = 48.0$, $Z = -20.4$.

61. The Distance to the Galactic Center.—It appears to be a tenable hypothesis that the supersystem of globular clusters is coextensive with the Galaxy itself. Researches on variable stars in the Milky Way[4] will eventually afford an instructive check on this hypothesis. Until we have such direct measures we can only assume that the galactic system is at least as large as the system of globular clusters, and note that we have shown rather convincingly that the globular clusters are galactic members or associates.

The algebraic mean of the values of $R \cos (\lambda - 327°) \cos \beta$ gives a satisfactory indication of the distance to the center of the cluster system and provisionally, therefore, of the distance to the center of the Galaxy. In so far as it depends on

[4] *Ibid*, H. Repr. 51, 1928.

the globular clusters now known, the uncertainty of the distance does not exceed ten per cent. Further research on faint globular clusters, especially if new ones be found, will be more likely to extend the system than to reduce it; on the other hand, the distances of the more remote clusters are the least certainly determined and must be given low weight. We shall adopt as the distance to the center

$$R_g = 16 \text{ kiloparsecs}$$
$$= 52,000 \text{ light years}$$

The galactic star clusters and the ordinary individual stars are too near the sun to contribute effectively to the determination of the distance to the center; but the direction to the center, as is well known, is confirmed by the counts of faint stars[5] and through the recent studies of galactic rotation by Oort, J. S. Plaskett, Lindblad, Schilt, and others, it is shown, though less definitely, by the distribution of Milky Way star clouds, planetary nebulae, Class O stars, and other objects of high luminosity. Most galactic objects, however, with the possible exception of the novae and Cepheid variable stars, are too faint absolutely and too infrequent in number to contribute in current surveys of galactic regions at and beyond the center of the cluster system.

Studies of the cluster-type Cepheids and the long-period variables stars in the star clouds of the southern Milky Way have led the writer and Miss Swope to a value of the distance of the centrally located star clouds that is very much like the value given above.[6] The variable star investigations of several observers at Harvard tend to support this suggestion that the heavy star clouds in Ophiuchus, Sagittarius, Scorpio, and neighboring constellations are parts of a massive stellar nucleus of the galactic system. The distribution of the cluster-type Cepheids in these regions suggests that the nucleus extends perhaps halfway from the center toward the sun. The speed

[5] Nort, Recherches Ast. Obs. Utrecht, 7, 113, 1917; Seares, Mt. W. Contr. 347, 1927.
[6] Shapley, H. Repr. 52, 1928.

of rotation about this nucleus is approximately 300 kilometers a second at the sun's distance from the center.

One important feature of the galactic central region is that the center itself lies behind heavily obscuring nebulosity. The dark clouds are apparently but a part of those causing the rift in the Milky Way that extends from Cygnus southward to Centaurus; they seem to be largely responsible for the apparent avoidance of the mid-galactic regions by globular clusters.

62. Galactic Dimensions.—An inspection of Figures XII, 1, XII, 2, and XII, 3 gives only a rough idea of the total diameter of the flattened galactic system. The most remote globular clusters are the following:

N. G. C.	Distance in Kiloparsecs	N. G. C.	Distance in Kiloparsecs
6325	46:	6517	50:
6342	40:	6528	44 4
6356	50:	6864	48 5
6440	50:	7006	56 8
6453	50:		

Some of these values are uncertain; the distances may be greater or less; but the superior distance of N. G. C 7006, a cluster in Vulpecula with $R = 56.8$ kiloparsecs = 185,000 light years, seems to be well attested by observations on its individual stars and on its cluster-type Cepheids.[7]

The greatest distance separating two of the clusters is

N. G. C. 7006 to N. G. C. 2298 = 80 kiloparsecs

Several other clusters are nearly as widely separated; N. G. C. 2298 in Puppis is 71.3 kiloparsecs from N. G. C. 6517 across the sky in Ophiuchus, and N. G. C. 2419 in Lynx is 79 kiloparsecs from N. G. C. 6453 in Scorpio. We may take such distances to indicate the extreme dimensions of the galactic system, for certainly most of these globular clusters, if not all, are integral parts of the Galaxy.

It does not follow from the wide dispersion of globular clusters that individual stars of the Galaxy are so widely dispersed, but

[7] Shapley and Mayberry, Mt. W. Comm. 74, 1921.

it appears reasonable to maintain that the greatest diameter of the Galaxy in its plane is not less than 70,000 parsecs, and it may be 30 per cent larger. The thickness of the system differs with distance from its center, being perhaps 10,000 to 15,000 parsecs at the galactic nucleus and one half as much out where the solar system is located. Occasional isolated stars, however, and, of course, the globular clusters extend to 20,000 parsecs and more from the galactic plane, lying well outside the relatively thin mid-galactic stratum.

Various studies of the extra-galactic nebulae, carried on, for the most part, at Mount Wilson, Upsala, and Harvard, seem to leave no doubt but that our galactic system is extremely large compared with typical external systems. The Andromeda nebula may have one fifth the diameter, but such gigantic spirals appear to be rare. Among the hundreds of thousands, possibly millions, of discoverable external systems or island universes in the oceans of space, our own system tends to be continental in dimensions. Whether it is comparable in form and structural detail with typical spirals, or more analogous with the irregular Magellanic Clouds, is a matter for the researches of the immediate future. Extreme irregularity of stellar distribution and heterogeneity of internal motion appear to characterize the Galaxy; but such irregularities may be normal in the arms of a great spiral, for all we now know, and are therefore not conclusive evidence that the Galaxy is a sheet of intermingling star clouds. In Chapter XIV this subject is again discussed and a revised hypothesis of galactic structure is proposed.

63. The System of Galactic Clusters.—Although the loose star groups contribute as little as individual stars of ordinary type to our knowledge of the dimensions of the Galaxy, their space distribution may be mentioned here as bearing on the structure of the nearer parts of the Milky Way. The catalogue of 249 galactic clusters in Appendix B gives galactic longitude, distance, and $R \sin \beta$ (distance from the galactic

plane); these quantities are used in making the plots for Figures
XII, 4 and XII, 5, which supplement the figures in Chapter II
dealing with apparent distribution.

Features to be noted in the diagrams include:

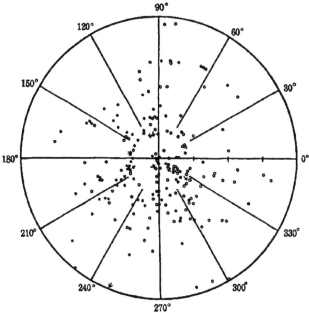

FIGURE XII, 4

Distribution of galactic clusters in the galactic plane within four kiloparsecs of the sun (origin of coordinates, which are galactic longitude
and distance from the sun). Dots symbolize clusters north of the
galactic plane; circles, those south. Arrows represent clusters beyond
the limits and in the directions indicated. For clusters with galactic
latitude β greater than $\pm 20°$, the projection on the plane $R \cos \beta$ is
plotted instead of the distance R. The radial scale in kiloparsecs is
marked along the zero axis.

1. The contrast in galactic distribution of galactic and globular clusters (see Figure II, 5).

2. Close confinement of galactic clusters to the neighborhood
of the galactic plane.

3. The relative nearness of galactic clusters to the sun, which
results in a distribution free from effects of the obstructing

clouds that contribute much to the anomalous distribution of the globular clusters.

The nearest globular clusters are:

N G. C.	Name	Distance in Light Years	$R \sin \beta$ in Light Year
104	47 Tucanae	22200	−15600
5139	ω Centauri	22200	+ 5900
6121	Messier 4	23500	+ 6200
6397	Dunlop 366	18400	− 3900
6656	Messier 22	22200	− 3600

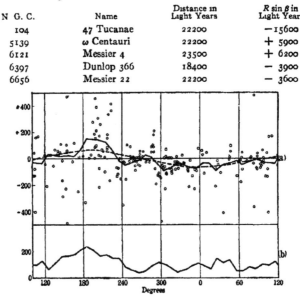

FIGURE XII, 5.

Distribution of galactic clusters. (a) Galactic longitude (abscissae) and $R \sin \beta$ (in parsecs), full line indicates 10° algebraic mean $R \sin \beta$, smoothed over 30°. (b) Smoothed 10° arithmetic means for same material, coordinates as for (a). Three clusters are off at upper left.

The most remote galactic clusters, excluding those for which the data are uncertain, are the following.

N G. C.	Distance in Light Years	$R \sin \beta$ in Light Years
2236	27200	− 200
2259	51600	+3200
2324	22600	+1900
H8	21600	−1800
H9	21600	+ 400
6005	21600	−1500

It seems clear that larger telescopes will eventually reveal many galactic clusters in remote parts of the Galaxy.

The most remote individual stars known in the Galaxy are the novae and a few long-period Cepheid variables, of faint apparent magnitude, studied by Gerasimovič, Miss Swope, Miss Harwood, and others on Harvard and Nantucket plates. Some of these appear to be well beyond the galactic nucleus, with distances comparable with those of the remoter globular clusters and the Magellanic Clouds. Since progress is rapid in the discovery and study of faint variables, it will be advisable to postpone the discussion of the part they play in the measurement of the extent and orientation of the galactic system. It may be 20 years or more before the detailed picture of the structure of the galactic star clouds can be drawn.

CHAPTER XIII

STAR CLUSTERS IN THE MAGELLANIC CLOUDS

A PECULIAR importance attaches to the two clouds of Magellan. They are near enough to be completely resolved into millions of stars, remote enough to be viewed and worked objectively, and rich in the various types of stars and nebulae. They serve as keys to knowledge of distant galaxies, opening the way from local regions to the outside universe.

Investigation into the structure and content of the Magellanic Clouds will assist, first, in the interpretation of the Milky Way clouds and the galactic system as a whole and, secondly, in revealing the nature of the more distant and inaccessible star clouds, such as N. G. C. 6822 and others of the Magellanic type of extra-galactic nebula. Probably two or three per cent of all recorded external systems are of this irregular kind.

In the Magellanic Clouds there are clusters of both the globular and the galactic type, and among the latter there is much variety in richness, dimensions, and nebulosity. For both types we can determine with some exactness the range in linear dimensions and in real luminosities. The distances from the observer can be assumed the same throughout each cloud; the apparent magnitudes, therefore, differ from absolute magnitudes only by an additive constant, and angular diameters from linear diameters only by a simple factor.

64. A Summary of Clusters and Nebulae.—In the New General Catalogue and the Index Catalogues 41 clusters and nebulae are listed within the limits of the Small Cloud and 301 within the Large Cloud. The descriptions, which are based mainly on Sir John Herschel's visual observations of about a century ago, are meager. Photographic plates show many of the older descriptions to be quite inadequate and reveal scores of clusters and nebulae that have not yet been catalogued and described.

For the Small Cloud a catalogue by Shapley and Miss Wilson[1] gives 237 new objects, chiefly nebulous stars and groups of stars. The Harvard photographs are incapable of differentiating clearly the nebulae or clusters that are fainter than the fourteenth magnitude; objects of these classes that are absolutely fainter than magnitude -3 are therefore as yet unlisted. A manuscript catalogue of the star clusters in the Large Magellanic Cloud, recently prepared from Harvard plates, contains 110 new entries—nearly doubling the number known from Herschel's survey and subsequent investigations. No attempt has yet been made to list the many nebulae that do not appear in published catalogues.

Among the great number of nebulae in both Clouds, not a single object has been found which could be assigned safely to the spiral class. As would be expected from their total absolute magnitudes, mainly between -3 and -7, the nebulae are, for the most part, of the diffuse and irregular types, with a meager sprinkling of planetaries.

In the Large Cloud many of the diffuse nebulosities are of exceedingly high total luminosity, the greatest of them being the famous "Looped Nebula," 30 Doradus. It is the largest object of its class, except possibly for some of the diffuse nebulosities connected with spiral nebulae; its absolute brightness probably exceeds magnitude -13, and its total diameter, including the fainter wisps and loops of nebulosity, is 30 parsecs or more. Many faint stars are involved, but apparently most of the registered luminosity comes from the nebula itself. It differs in this respect from the majority of the diffuse nebulae in the Clouds; their apparently excessive luminosity is largely that of involved bright stars or star clusters.

Compared with 30 Doradus the great Orion Nebula is a pigmy. If the former were in the position of the Orion Nebula,

[1] H. C. 275, 276, 1925.

it would extend over most of the constellation of Orion, and its brightness would be so great that, even though 600 light years distant, it would cast strong shadows on the surface of the earth.[2]

To catalogue and describe in detail the numerous clusters of the Magellanic Clouds is inadvisable at the present time in view of the new material that will soon be available. Heretofore the instruments mainly responsible for our knowledge of the structure of the Clouds have been the Lick Observatory spectrograph at Santiago, Chile, which has given us the radial velocities of bright line nebulae, and the Bruce 24-inch doublet at the Boyden Station of the Harvard Observatory. On plates of suitable exposure the Bruce telescope shows objects of the eighteenth magnitude and fainter, but the sacle ($1' = 1$ mm.) does not permit adequate resolution of the small clusters. The new 60-inch reflector of the Harvard Observatory will soon come into use in the comprehensive study of the Magellanic Clouds. Until its surveys have made considerable progress, we must not only leave open the cataloguing and analysis of the clusters and nebulae but also be content with somewhat provisional values of the distances and dimensions of the Clouds themselves.

65. The Globular Star Clusters.—Dunlop, Sir John Herschel, and others have described many of the compact star groups in both Clouds as globular. The N. G. C. records 16 globular clusters in the Large Cloud and two in the Small. On the basis of photographic material, Bailey, Melotte, and others have remarked that few if any of these objects are correctly assigned. In fact, none of them is now retained as truly globular. On the other hand, the accepted globular clusters, listed for both Clouds in Table XIII, I, are not described as such in the N. G. C., though they all appear in that catalogue. The

[2] Shapley and Wilson, H. C. 271, 1925. Allowance has been made in the foregoing statements for the revision of the distance of the Large Magellanic Cloud. A final determination has not been made, but it appears best to reduce by half a magnitude the absolute magnitudes of clusters and nebulae previously published.

existence of these globular systems affords a valuable opportunity for the comparative study of globular and galactic types
and for the examination of the relation of globular clusters to
galaxies.

The first two clusters of Table XIII, I belong to the Small
Magellanic Cloud; the eight others to the Large Cloud. The
angular diameters and integrated magnitudes are given on the
same basis as in Appendix A for globular clusters in general.
The diameters, therefore, are not indicators of the extreme
bounds of the clusters—they are rather measures of nuclei.
N. G. C. 416 in the Small Cloud is more uncertain than the
others and later may be dropped.

It is seen that the globular clusters in the Large Cloud range
from the compact Class II to the fairly open Class VII. In
earlier considerations of clusters in the Large Cloud[3] seven
objects were listed as possibly globular. Of these N. G. C.
1651 has now been definitely dropped,[4] and N. G. C. 1835
and N. G. C. 1856 have been added to the list. Later analysis
may show that some of the following N. G. C. objects are
globular clusters:

1711	1916	1986	2058	2133
1789	1926	2019	2065	2134
1852	1939	2031	2107	2157
1872	1944	2056	2108	2164
1903				

In none of the globular clusters of the Magellanic Clouds
have variable stars been found, nor have their brighter stars
been measured individually. The final test as to whether these
doubtful objects are typical globular systems or merely open
groups involved in nebulosity will lie in future examinations
for variable stars and in the study of density and luminosity
laws.

The spectrum of N. G. C. 419, in the Small Cloud, resembles
Class K in its distribution of light and in the faint appearance

[3] H. B. 775, 1922; H. C. 271, 1925.
[4] It was included in H. B. 848, 849, and 852 as a doubtful object.

TABLE XIII, I.—GLOBULAR CLUSTERS IN THE MAGELLANIC CLOUDS

N. G. C.	R. A. 1900	Dec 1900	Galactic Long	Galactic Lat	Angular Diameter	Integrated Magnitude	Class	Adopted Modulus	Quality
416	1 5.0	−72 53.5	262	−43	0 9	11 3	VI	18 05	e
419	1 5.4	−73 25.1	262	−42	1 4	10 2	IV	17 36	c
Mean		.	.					17 50	
1783	4 58.8	−66 8.1	242	−37	1 4	10 1	VII	17 32	c
1806	5 2.4	−68 8.0	245	−36	0 9	10 6	VI	17 94	d
1831	5 5.8	−65 3.6	242	−35	1 3	10 0	V	17 38	c
1835	5 5.8	−69 32.1	247	−34	1 2	9 9	II	17 44	d
1846	5 7.7	−67 35.0	245	−34	1 2	10 4	VIII	17 56	c
1856	5 10.1	−69 15.1	247	−34	2 1	8 8	V	16 70	c
1866	5 13.6	−65 34.9	241	−34	2 2	8 0	IV	16 58	c
1978	5 28.1	−66 18.5	242	−33	1 0	10 2	VI	17 72	d
Means	:	:.....			1 5	9 64		17 25	
						±0 24 (m e)		±0 11 (m e)	

of the G band and of lines H and K. The spectrum of N. G. C. 416 is too diffuse and faint to classify. The spectral class[5] of N. G. C. 1866, in the Large Cloud, is F8; that of N. G. C. 1835 is possibly G5, though for it and the other most compact clusters of the Large Cloud the spectral images on the Harvard objective prism plates are too difficult for classification.

Miss Cannon has thrown doubt on the globular nature of a number of bright objects in both clouds by showing that their integrated spectra are of early class; we have already seen that practically all typical globular clusters belong to classes F and G.[6] Thus, she finds:

N. G. C.	Spectrum
294	A?
1872	A3
1903	A
2041	A3
2107	A?
2134	A
2157	A2
2164	A5

66. Distances of the Clouds Derived from Variables and Globular Clusters.

—In the ninth column of Table XIII, I a value of the distance modulus is given for each cluster, derived from the measures of angular diameter and integrated photographic magnitude. In the use of these data we follow the principles developed in an earlier chapter on the distances of globular clusters of the galactic system. In getting mean values of the modulus for each Cloud, weights were assigned as follows:

Quality	Weight
c	4
d	2
e	1

While clusters of quality c may be accepted as almost certainly globular, some doubt still attaches to those qualified as d and e.

[5] H. B. 868, 1929.
[6] See Appendix A.

The mean modulus derived for the Small Magellanic Cloud is nearly identical with the value, 17.55, previously derived from 107 Cepheid variable stars.[7] With the adopted revision of the zero point of the period-luminosity curve[8] the distance modulus from the variable stars becomes 17.32. Accepting this value, we have for the Small Magellanic Cloud

$$\pi = 0''.0000345$$
$$\text{Distance} = 29 \text{ kiloparsecs}$$
$$= 95,000 \text{ light years}$$
$$\text{Linear diameter} = 6,000 \text{ light years}$$

At this distance the integrated absolute magnitude is -7.12 for N. G. C. 419, the only object that appears definitely, from the survey of existing plates, to be a globular cluster.

For the Large Magellanic Cloud the mean distance modulus from eight globular clusters is 17.25, with a mean computed error of only one tenth of a magnitude. But the Cepheid variable stars should eventually give us a much more dependable value of the modulus. The preliminary determination,[9] giving a modulus from variables of approximately 17.7, was based on few stars and somewhat provisional magnitude sequences. The sequences are not yet wholly satisfactory, but a current unpublished study at Harvard has produced periods of about 50 Cepheid variable stars in the Cloud and yields a value of 17.10 for the distance modulus. Adopting this value, we obtain the following results for the Large Magellanic Cloud:

$$\pi = 0''.000038$$
$$\text{Distance} = 26.2 \text{ kiloparsecs}$$
$$= 86,000 \text{ light years}$$
$$\text{Linear diameter} = 10,800 \text{ light years}$$

The corresponding mean absolute photographic magnitude of a

[7] Shapley, Yamamoto, and Wilson, H. C. 280, 1925; see, also, Shapley, H. C. 255, 1924.

[8] See Chapter X.

[9] Shapley, H. C. 268, 1924.

globular cluster in the Large Magellanic Cloud is -7.46 (weighted mean). The absolute values range from -9.1 to -6.5—an indication of the degree of uncertainty involved in assuming a constant integrated absolute magnitude for globular clusters. The spread is considerably less if the cluster N. G. C. 1866 is assigned to the foreground rather than to the Magellanic Cloud, and it would be very small indeed (-7.2 to -6.5) if the newly admitted N. G. C. 1856, when analyzed with a large telescope, proved to be a nebulous open group.

67. On the Relation of the Clusters to the Magellanic Clouds.—The angular diameters of the globular clusters in the Large Cloud are at most two or three minutes of arc; the angular diameter of the Cloud as a whole is slightly more than seven degrees.[10] Enormous and rich as we know a typical globular cluster to be, it is obviously small compared with ordinary external galaxies. The clusters of the various sorts, however, are important in the general appearance of the Clouds and especially in the make-up of the high-luminosity population.

The distribution of the nebulous groups throughout the Clouds is much the same as the distribution of the Cepheid variable stars and of the general stellar population; on the other hand, the accepted globular clusters in the Large Cloud are almost exclusively to the north, and N. G. C. 1866 and N. G. C. 1831 lie quite outside the main structure of the Cloud. There are, however, a half-dozen variable stars, apparently of the Cepheid class, in the same region, and long exposures on small-scale plates show that these outlying clusters and variables are within the observable bounds of the Cloud.

The asymmetrical distribution of the globular clusters in the Large Cloud has led some to surmise that the globular clusters of the galactic system may also be eccentrically arranged with respect to the general galactic structure and, therefore, that they cannot be used, as I have used them, in estimating galactic dimensions. But this one-sided distribution of the globular

[10] *Ibid.*

clusters in the Large Cloud is modified by the inclusion of
N. G. C. 1835 and N. G. C. 1856 in the list accepted at present,
and it would be quite altered if a considerable proportion of the
list of suspected globular clusters prove to be typical systems.
The two clusters N. G. C. 1789 and N. G. C. 1944, both of
which are strongly suspected as globular, lie far from the center
of the Cloud on the south—directly across from the outlying
globular clusters of Table XIII, I.

It is of significance that the 808 known variable stars in the
Large Cloud and the 969 in the Small Cloud[11] avoid the numerous open clusters. In this respect the Clouds are like the
galactic system, where open clusters are free of variable stars
of all kinds.[12] The globular systems in the Magellanic Clouds
are, of course, too compact to have been searched as yet for
variables.

Possibly the most striking fact arising from the study of
globular clusters in the Magellanic Clouds is the low absolute
magnitude of their brightest stars when compared with the
brightest objects in the open and nebulous groups. If the new
values of the distance moduli are correct, there is scarcely a
star in the ten accepted globular systems of the two Clouds
that exceeds −3 in absolute photographic magnitude. This
result, however, is in complete agreement with the condition
found in globular clusters of our own galactic system. In
contrast, the individual stars in scores of the nebulous groups
appear to exceed −3 in absolute photographic magnitude,
and in many they attain the excessively high luminosities of
−5 and −6. This again is in agreement with the data on
absolute magnitudes in galactic clusters, especially in those
early-spectrum groups in Orion, Scorpio, and elsewhere that
are associated with bright nebulosity.

It may be remarked in conclusion that, notwithstanding their
remoteness, isolation, and high galactic latitudes, the Magellanic
Clouds may be more closely allied to our galactic system than

---

[11] Leavitt, H. A., **60**, No. 4, 1908.
[12] See Chapter IV, Section 20.

heretofore suggested.[13]   There is growing evidence for galactic rotation, and from a preliminary value of its velocity we compute that the observed high-speed recession of the Clouds should be assigned almost wholly to the rotation of the Galaxy. Apparently, the Clouds are not in rapid motion with respect to the galactic nucleus or the system of globular clusters.

[13] H. Repr. 61, 1929.

CHAPTER XIV

DATA BEARING ON THE ORIGIN OF THE GALAXY

It is encouraging to see how fragile and futile are the majority of astronomical theories and speculations and how temporary are the most conservative "interpretations"; for the futility of speculations emphasizes the importance and durability of observations and indicates the steady progress of the science. Nevertheless, interpretations must be ventured in order to fill in the picture where observations do not or cannot touch, and also to guide further measurement and analysis.

It was not unexpected that the working hypothesis of the origin and growth of the galactic system, suggested several years ago as an aid in the study of the observations then available on clusters, nebulae, and star clouds, would soon require extension and repair. The current researches on extra-galactic nebulae and on the variables in the Milky Way contribute vigorously now and will add still more in the near future, to the theory of the Galaxy and comparable systems. Meanwhile, various data bearing on galactic origin and structure can be assembled in this chapter and tentatively appraised; in the final section a modified hypothesis of galactic structure is advanced.

68. The Earlier Interpretation.—In the discussions of galactic origin and behavior that grew out of the earlier work on clusters and the local star system,[1] it was suggested that the discoidal galactic system, originating from the combination of independent star clouds and clusters, has long been growing, by assimilating such groups, to its present relatively enormous size. Its diameter was found to be of the order of 90,000

---

[1] Mt. W. Contr. 157, Sec. 7, 1918.

parsecs, and the center was located in the remote star clouds of the southern Milky Way. Evidence was then found (1918) that the galactic system is moving with high speed through space, with respect to the brighter extra-galactic nebulae, toward some ill-defined point high in the northern hemisphere.[2] Subsequent investigations of this motion by Wirtz, Lundmark, Strömberg, and others confirmed the general direction, the speed, and the uncertainty of apex.

The moving Galaxy was visualized as collecting in the course of time subordinate external systems and gradually dismembering and absorbing them. The local cloud was mapped out and described as one of these minor elements, though previously it had been treated as the major part. The Galaxy was considered not a single spiral nebula but rather an organization of many half-digested star clouds and clusters, moving in an extensive stratum of stars and galactic nebulae.

It was suggested that the obvious dynamical equilibrium of a globular cluster, acquired originally at a great distance from external perturbing matter, results in a delicate adjustment that readily breaks down under stresses such as those prevailing in a large galactic system; further, that faint stars in a globular cluster, as in the galactic system, are of small mass[3] and, therefore, of more than average velocity, so that in their motions in the cluster they frequently attain great distances from the center. When such globular systems approach or mingle with other clusters or the dense stellar fields in the mid-galactic segment, the dwarfs will preferentially become scattered through encounters. The massive cluster stars, which are mostly of high luminosity, and which are concentrated to the center and endowed with low peculiar velocities, will retain their cluster organization longer in a disrupting neighborhood. A globular cluster thus becomes a galactic cluster which slowly dissolves into the galactic field. Flattening and distortion of

---

[2] *Ibid.* 161, Sec. VII, 1919.

[3] In view of subsequent work on the mass-luminosity relation there is now little of the hypothetical about this suggestion.

galactic clusters and star clouds arise through the "encounter machinery" discussed by Jeans and through rotation. Star streaming appears to be a complication of the various motions in and of the local system and the general galactic field.

On the provisional interpretation sketched above, our galactic system appears to be in an advanced stage of the survival of the most massive. Once an enormous mass has accumulated in such a celestial organization, subsequent accretions should be numerous and relatively easily acquired. The Magellanic Clouds (and we should now include their analogues among the extra-galactic nebulae) are possibly similar growths. The Clouds, however, are relatively so near and are so much smaller than the Galaxy in mass and dimensions that the possibility of ultimate assimilation into the galactic fields is not excluded, notwithstanding their present large positive velocities in the line of sight; they are, indeed, not receding rapidly if we correct for "galactic rotation."

In earlier work on the galactic problem, I found it necessary to conclude that the spiral nebulae are not comparable to our Galaxy in size. The conclusion was based largely on observations of novae. Assuming the novae in spirals to be comparable in luminosity to those in the Galaxy, in 1917 I computed the distance of the Andromeda Nebula[4] to be approximately a million light years; but even at this great distance (rather an overestimate) it was shown to be much smaller than our Galaxy. The lack of comparability between galactic system and spiral nebula appears now more certain than before; ours is a Continent Universe if the average spirals are considered Island Universes. In view of this incomparability in dimensions and mass, and especially because of the evidence for proper motions in the spiral arms, I temporarily abandoned the hypothesis that the nebulae of the spiral family are stellar in composition. But the stellar composition of at least some of them has now been proved; the arguments against the Kant-Herschel theory of a plurality of universes have gone down under the weight of

[4] P. A. S. P., 29, 216, 1917.

novae and variable stars. The two results that seem most convincing in establishing the stellar character of the spiral family of nebulae are Lundmark's observation of the generality of the Magellanic type of star cloud among the extra-galactic nebulae[5] and Hubble's work on the Cepheid variables in several typical bright spirals.

69. The Research on Milky Way Variable Stars.—The measurement of the size of the galactic system, discussed in Chapter XII, depends primarily on the Cepheid variable stars and other stars of high luminosity in globular clusters. It is unsatisfactory to continue to base estimates wholly on the distribution of globular clusters. Direct measurements, however, are difficult for two reasons: (1) there is interference by obscuring clouds in low galactic latitudes, and (2) extensive labor is necessary to map out the irregular star clouds and the galactic boundaries by determining the distances of the stars one by one.

The customary statistical method of deducing details of galactic structure from rather indiscriminate counts of stars and from measures of motions in the solar neighborhood is very limited in scope and is, in fact, wholly inadequate for analyses of regions characterized by galactic star clouds some 20,000 to 100,000 light years distant. The direct attack by those photometric methods that reach far and give unambiguous results on the problem appears to be the only satisfactory way of working out details of galactic dimensions and structure. In time, spectroscopic parallaxes and spectral parallaxes of distant stars may be highly effective, but at present the best approach seems to be through variable stars of all types.

To meet the need for more information on faint variable stars, a long program of variable star investigations was inaugurated in 1923 at the Harvard Observatory. The program has

[5] Pop Astro. Tidsk., **7**, 64, 1926. He lists 21 systems as probably being of the Magellanic type. At Mount Wilson and Harvard it has been pointed out that two or three per cent of all external systems are irregular in form and are probably star clouds (see Chapter XIII).

been described elsewhere;[6] it will suffice here to mention briefly a few of the results that already begin to throw light on the origin of the Galaxy.

1. Since the starting of the variable star program, about 2,500 new variable stars have been discovered on Harvard plates, nearly doubling the number previously known in the galactic system.   The variables are largely of the Cepheid and long-period classes; they are accordingly giants.   For more than half, the maximum magnitudes are fainter than 14, and they are therefore at least 20,000 light years distant.

2. Indications of a massive nucleus of the Galaxy in the Sagittarius region have already been described in Section 61, above.[7]

3. The diameter of the galactic nucleus, if we judge by the variable star data and the long-exposure photographs of the Milky Way, is something like 35,000 light years.

4. Long-period variable stars are found to be numerous in Milky Way Field 185 and adjacent regions; we have, therefore, hopes that this type of variable can also be calibrated and that it will prove useful in estimating the distances of star clouds.

5. The distance to the Sagittarius clouds is practically the same as the distance derived from the globular star clusters for the center of the galactic system.   It appears reasonable to assume tentatively that these star clouds actually mark the center of the larger galactic system and that it is their total mass that is important in the motions of our local system.

70. Peculiarities in the Distribution of Galactic Clusters.—The clusters listed in the catalogue in Appendix B do not extend deep into the galactic structure; they tell us nothing of the center or of the boundaries.   There is significance, however, in certain peculiarities of their distribution.   Already we have noted that in contrast with globular clusters they are rather uniformly dispersed in galactic longitude, and also that

---

[6] Shapley H. Repr. 51, 1928.
[7] See, also, H. Repr. 52, 1928.

they are largely confined to the low galactic regions where globular clusters are scarce. There are two additional features of their distribution that are worth consideration:

1. The infrequency of galactic clusters in the first quadrant of galactic longitude; there are very few of these systems in the Aquila-Cygnus region of the Milky Way.

2. The narrow restriction of the clusters to low galactic latitudes in the direction of the galactic center and their wide dispersion in galactic latitude in the opposite part of the sky.

This second phenomenon is more conspicuous in the second figure of Chapter II, in which ordinates are galactic latitudes, than in the fifth figure of Chapter XII, in which the ordinates are distances from the galactic plane. The distribution might be explained as a consequence of the motions of galactic clusters in long orbits about the nucleus in Sagittarius. When such objects are seen from the earth's eccentric position in the Galaxy, those in the direction of the center would appear to be in lower latitudes than those away from the center, except when the orbits lie exactly in the galactic plane. It is probable that none of the galactic clusters beyond the center of the Galaxy enters our catalogue.

The alternative and, I think, preferable interpretation is that galactic star clusters are associated largely with particular galactic star clouds. Irregularities of distribution and concentration in low galactic latitude are therefore merely consequences of the distribution of such star clouds as the local system in this part of the Galaxy. If the galactic clusters are closely affiliated with various star clouds, we may find them differing systematically in spectral and structural characteristics from one part of the sky to another. We already have an indication of such diversity in the orientation of the axes of elongation, discussed in Section 33. The bright clusters of Auriga (Messier 36, Messier 37, Messier 38) are rich and not strongly condensed; many of the small condensed clusters of Sagittarius are nebulous, and the groups in Carina are systematically bright.

71. Radial Velocities of Globular Clusters.—The measured radial velocities of globular clusters range from −350 to +315 kilometers a second. From Table XIV,I it can be seen that there is no clear dependence of velocity on class of

TABLE XIV, I.—RADIAL VELOCITIES OF GLOBULAR CLUSTERS

N. G. C	Class	β	Spectrum	Angular Diameter	Pg Mag.	Distance	Radial Velocity
				′		k pc	
1851	II	−34 5	..	5 3	6 0	14 3	+315
1904	V	−28	...	3 2	8 1	20 4	+235
5024	V	+79	.	3 3	6 9	18 2	−180
5272	VI	+77 5	G	9 8	4 5	12 2	−130
5904	V	+46	G	12 7	3 6	10 8	+ 10
6093	II	+18	KO	3 3	6 8	17 5	+ 70
6205	V	+40	GO	10 0	4 0	10 3	−265
6218	IX	+25	..	9 3	6 0	11 0	+160
6229	VII.	+40	.	1. 2	9 7	29 8	−100·
6266	IV	+ 7	KO	4 3	7 0	18 6	+ 50
6273	VIII	+ 9	G5·	4 3	6 8	16 3	+ 30
6333	VIII	+10	K?	2 4	7 4	20 8	+225
6341	IV	+35	G5·	8 3	5 1	11 2	−160
6626	IV	− 7	G5	4 7	6 8	16 6	0
6934	VIII	−20	GO	1 5	9 4	24 9	−350
7078	IV	−28	F	7 4	5 2	13 1	− 94
7089	II	−36	F5	8 2	5 0	13 9	− 10
7099	V	−48 5	I·8	5 7	6 4	14 6	−125

cluster, galactic latitude, angular diameter, total magnitude, or distance from the sun. The only appreciable correlation appears to be that of speed with distance from the solar apex, pointed out by Strömberg.[8] The dependence is in the sense of increasing velocity with increasing distance from the solar apex.

[8] Mt. W. Contr. 292, 5, 1925. A suggestion of a dependence of velocity (corrected for solar motion) on galactic latitude and therefore on mass of the intervening star fields is discussed by ten Bruggencate (P. N. A. S., 16, 111, 1930), who seeks a trace of the red-shift, characteristic of the spectra of distant spiral nebulae. The material is as yet insufficient to establish the correlation securely or to discriminate among its possible interpretations.

The simplest interpretation of the relation of speed to position in the sky is that the apparent systematic drift of the clusters is but the reflection of the motion of the local system in the Galaxy. When corrected for this motion, the average speed remains high—approximately 100 kilometers a second; but as a group, the globular clusters are essentially at rest, unlike the extra-galactic nebulae, which show a large K term apparently dependent on distance.

The radial velocities of globular clusters have been measured mainly by Slipher at the Lowell Observatory. Except for two or three clusters, all measures refer to the integrated images and not to individual stars. The study of differential radial motions in a globular cluster is one of our important future problems. The successful measure of the proper motions in globular clusters also must await the photographs of the future. Van Maanen has shown that the proper motion of Messier 13 as a whole and its average internal proper motion are each less than 0''.001 annually, an amount to be expected from a consideration of the distance and the radial velocity.[9] The values of the annual proper motions are slightly larger for Messier 2 and Messier 56 but are consistent, he finds, with my estimated distances and the average radial velocities.[10]

72. Dimensions and Star Densities of Clusters.—In the earlier hypotheses concerning the origin and growth of the Galaxy, it was suggested that the galactic clusters probably represent stages in the dissolution of a typical globular system. The existing galactic clusters are necessarily in the process of dispersion through encounters; but it is not equally evident that existing globular clusters are doomed to assimilation by the Galaxy and subsequent transformation. We have already commented in Section 55 on the giant-poor clusters, and in Section 8 we have called particular attention to N. G. C. 2477, one of the richest and most globular-like of the accepted galactic

[9] *Ibid.* 284, 1925.
[10] *Ibid.* 338, 1927.

clusters. Such transition types are, however, apparently scarce, and in considering the possible relationships of the two groups, it is well to examine further their comparative dimensions and star densities.

It is impossible to say how many stars constitute a typical globular cluster. Our photographs can reach only a little way down the main sequence toward the dwarfs. When we attempt to go farther, the high density of the central stars "burns out" the photograph and conceals the information we might otherwise obtain. From available counts on our most suitable photographs of the brightest clusters we estimate[11] that in the average globular cluster there are more than 20,000 stars brighter than absolute magnitude $+5$. To the same magnitude limit, the population of an average galactic cluster is less than 200 stars.

The diameter of a globular cluster is also indeterminate. It is probable that the actual linear dimensions depend on the brightness of the stars involved, becoming greater for stars of lower luminosity and mass; the same dependence appears also in galactic clusters.[12] From Table XI, III (Section 56), which gives the relation of distance modulus to angular diameter for normal globular clusters, we can compute the following relation of linear diameter to distance:

Distance	Modulus $m - M$	Angular Diameter	Linear Diameter
kpc		'	*pc*
10	15.0	11.2	33
20	16.5	2.6	16
30	17.4	1 4	13
40	18 0	1 02	12
50	18.5	0.85	12

The measured decrease of linear diameter with increasing distance is, of course, mainly photographic. The loss of light in space is here of minor significance. We under-measure the angular diameters of remote clusters because of the failure of outlying faint stars to rise to measurable prominence on the

[11] See, for example, Pop. Astr., **27**, 101, 1919.
[12] See Figure VII, 2.

photographic plate. I think we can safely take the diameter of a typical globular cluster to exceed 35 parsecs; but the diameter of the nucleus, in which the brightest stars are concentrated, appears to be only one third as large.

The average linear diameter of the galactic clusters for which definite estimates can be made (Appendix B) is 6.24 parsecs. Few galactic clusters exceed 20 parsecs in diameter. Trumpler has determined preliminary distances from observations of magnitudes and spectral types for 54 systems. He finds a range in linear diameter from 3.5 to 25 parsecs, with the great majority between 4.5 and 10 parsecs.

The number of stars per cubic parsec in a typical globular cluster cannot be computed at present, except for the supergiant stars. It is obvious, when the dwarfs are taken into consideration, that the distances separating stars at the center of a rich globular cluster are on a planetary rather than a stellar scale. It seems probable that sooner or later we should have evidence of stellar encounters in such crowded regions; but only one nova in a globular cluster is now on record—the seventh magnitude object in Messier 80, which appeared in 1860.

The space density and the distances separating individual stars can be more readily computed for galactic clusters when reliable estimates on the parallaxes become available and we have made sufficient allowance for superposed stars. Trumpler's study of one of the richest of the galactic clusters, Messier 11, provides material that illustrates the conditions in these systems. He finds that the cluster is 1,250 parsecs distant. It lies in the rich Scutum star cloud, which has a star density nearly four times that of the average field of the galactic belt. The cluster itself is made up of about 480 stars brighter than magnitude 15.5, distributed over an area approximately a quarter of a degree in diameter. The bright stars are concentrated within a central area of less than four minutes radius.[13]

[13] L. O. B., 12, 10, 1925. The distance is revised in his later work, to 1340 parsecs.

For stars brighter than absolute magnitude +4.5 the relation of density to distance in Messier 11 is found to be as follows:

Distance from Center in Parsecs	Stars per Cubic Parsec
0 27	83
0 60	80
0 96	33
1 32	9 5
1 68	4 9
2 04	2 2
2 40	0 5

The central density of Messier 11 is much higher than that of the average galactic cluster. In contrast with the density of 83 stars per cubic parsec for Messier 11 is that for Messier 37 of only 18 stars per cubic parsec for stars brighter than absolute magnitude +4.5. The corresponding number for Messier 36 is 12 (Wallenquist), for the Pleiades, 2.8 (if the parallax is taken as 0″.008), and, for the vicinity of the sun, 0.011. The average separation of stars at the center of Messier 11 is one light year. Trumpler points out that "an observer at the center of Messier 11 would find about 40 stars with parallaxes of 2″ or more and which would appear three to fifty times as brilliant as Sirius shines in our sky." It is quite probable, moreover, that this display would be very dull compared with the show at the center of the Hercules cluster.

73. On the Masses of Giant Stars.—The correlation tables for color index and apparent magnitude for three bright globular clusters are given in an earlier chapter. Using the observed mass-luminosity relation for galactic stars, we can transform the color-magnitude array into a relation between spectral class and mass. It is necessary to assume that we can safely replace color class by spectral class for giant stars; the uncertainties involved both in this assumption and in the temperature scale, used for reduction from photovisual to bolometric magnitude, are not negligible, but still they are not serious enough to falsify the average results except for stars of extreme color. It is possible that the chief source of error lies in the mass-luminosity curve itself, which depends mainly on nearby double stars and possibly is inappropriately applied to single stars, especially in a globular cluster.

The computations of mass have been carried through[14] for Messier 22. The color-magnitude array (Table III, V) includes the stars brighter than the magnitude limit of the photovisual plates and within five minutes of the center of the cluster. It is of interest that more than six per cent of the stars have negative color indices, the cluster resembling[15] in this respect Messier 13 rather than Messier 3.

No correction has been made in the color-magnitude array for superposed stars. The cluster lies in a rich star cloud in Sagittarius, and probably 10 per cent of the stars included in this discussion are not cluster members. The color-magnitude array is, therefore, applicable both to the cluster and to the star cloud, and the small dispersion in brightness of both together suggests that the two are associated. The computation of the stellar masses in terms of the sun's mass for successive intervals of magnitude and color is shown in Table XIV, II. The distance modulus $m_{pv} - M_{pv} = 14.16$ is taken from Appendix A; the reduction to bolometric magnitudes and the computation of the masses are made with the aid of tables given by Eddington.[16] The masses of the reddest stars would have been from 10 to 20 per cent less on Brill's scale of temperatures and corrections to bolometric magnitude.[17] The values of photovisual magnitude and spectrum in Table XIV, II are read directly from the curve drawn through the plot in Figure XIV, 1 of the colors and magnitudes of individual stars that appear in the color-magnitude array.

The following masses for the average giant stars of various spectral classes in Messier 22 are derived from the smooth curve

[14] H. B. 874, 1930.
[15] See Table III, II, Table III, III, and Mt. W. Contr. 155, 8, 1918.
[16] Internal Constitution of the Stars, Chapter VII, 1926.
[17] Babelsberg Veröff., 5, 16, 1924.

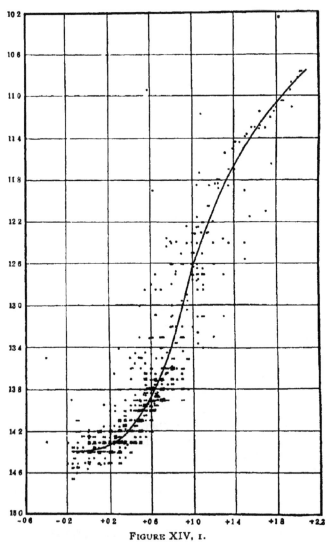

FIGURE XIV, 1.

Color-magnitude array for Messier 22.   Coordinates are photographic magnitudes and color indices.

in Figure XIV, 2, where the data are plotted from the second
and fifth columns of Table XIV, II:

Ao	<4 8	F5	4 o	Ko	10 8
A5	<3.5	Go	4 9	K5	16 5
Fo	<3 6	G5	7 o	Mo	24 o

TABLE XIV, II.—THE MASS-SPECTRUM RELATION FOR MESSIER 22

Apparent Pv Mag	Spectrum	Absolute Pv Mag	Absolute Bol Mag	Mass
11 2	M2 5	−2 96	−4 62	29
11 4	K9 o	−2 76	−4 10	22 4
11 6	K5 8	−2 56	−3 72	17 8
11 8	K3 o	−2 36	−3 13	14 1
12 0	K1 o	−2 16	−2 74	11 7
12 2	G9 o	−1 96	−2 41	10 0
12 4	G7 o	−1 76	−2 10	8 3
12 6	G5 5	−1 56	−1 84	7 4
12 8	G4 o	−1 36	−1 58	6 5
13 0	G2 5	−1 16	−1 30	5 9
13 2	G1 2	−o 96	−1 06	5 4
13 4	F9 8	−o 76	−o 80	4 8
13 6	F7 8	−o 56	−o 58	4 5
13 8	F5 8	−o 36	−o 36	4 1
14 0	F3 o	−o 16	−o 16	3 9
14 2	A9 5	+o o4	+o o4	3 7
14.3	A6 8	+o 14	+o o7	3 6
14 4	B7 5	+o 24	−o 49	5 7

It is possible to give only upper limits of average mass for
classes Ao, A5, and Fo because of the incompleteness of the
observational material for the corresponding intervals of color
index.

For the rich galactic cluster Messier 37, von Zeipel and
Lindgren, in good agreement with the present results, find the
mass of the giant g5 stars 2.15 times as large as the average
mass of the b and a stars.[18]   They have used space distribution
of the stars as a criterion and a measure of the masses for stars
of different types.[19]

[18] Proc. Swedish Acad., 61, No. 15, 126, 1921.
[19] See Chapter V, Section 26.

The most interesting feature of Figure XIV, 1 is the small dispersion in color for stars of a given photovisual magnitude. Accepting the mass-luminosity relation, we can only conclude that in a globular cluster such as Messier 22 the giant stars of a given mass have a very small spread in surface temperature.

A mass-spectrum relation, as deduced from Figure XIV, 1 and plotted in Figure XIV, 2, has heretofore never been derived for a globular cluster.

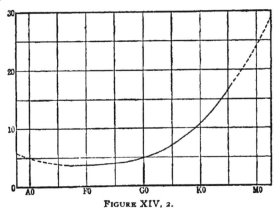

FIGURE XIV, 2.
Mass-spectrum curve for Messier 22.

74. On the Evolution of Globular and Galactic Groups.—

In the existence of supergiant stars in globular clusters there are some implications bearing directly on the principles of stellar evolution that have not as yet received the emphasis deserved. They bear indirectly on our problem of the development of the galactic system.

1. Apparently all of the globular clusters show the presence of numerous red supergiants, with the fainter stars always bluer. It follows that there must be for the typical globular system an "equilibrium in time"—an essentially stationary age, as remarkable as the observed dynamical equilibrium.

2. It would be unreasonable to assume exactly the same age for all globular clusters and argue therefrom that they are now similar in stage of development only because the same time

interval separates them from the date of origin. Even the different distances from the observer would, on account of the finite speed of light, give them effective ages differing by nearly 2×10^5 years. It appears more reasonable to admit that undisturbed globular systems are *effectively* permanent in spectral composition. Is this because the time units we use in measuring stellar evolution are still too anthropocentric or because the clusters are timeless?

3. No theory of evolution at present in vogue can explain supergiant stars that are stationary with respect to time—especially such exceptional objects as the low-density, high-luminosity red stars with normal spectral properties. The very act of radiation spells change; the loss of mass per second in a cluster giant exceeds a thousand million tons.

4. Whether it is the same star that always remains bright, massive (presumably), and red, or whether all cluster stars progress systematically and their places are methodically taken by others, we have as yet no way of knowing. It is, of course, unreasonable to ask that new supergiants be born at just the right rate to keep up appearances. It would be better either to assume essential permanency in the arrangement as now observed, thereby admitting that we have been much too terrestrial in trying to force on stellar processes a time scale easily conceivable to us; or to assume that stellar development is not unidirectional—that stars may, and perhaps often or always do, move up the luminosity sequence as well as down—toward gigantism as well as toward small mass and low luminosity. The part that meteoric matter plays in fueling the stars is an open research of much difficulty and some promise.

5. Time seems to leave its marks, however, on the clusters along the Milky Way. Some of the globular clusters may be affected in freedom, form, and eventual survival by contacts with galaxies or other clusters. The marks of age are shown best by the numerous disturbed galactic groups, with their variety in structure and content. But even the Hyades type

of galactic cluster, with its yellow and white giants side by side, leaves us again groping for rather far-fetched assumptions to disembarrass ourselves of the apparent unevenness in the rates of stellar evolution.

6. Finally, it is worth suggesting that we may have been using our theories of stellar evolution too much as explanations and not enough in their proper place as temporary guides.

75. The Galactic System as a Super-galaxy.—The relation of galactic and globular star clusters to each other and to star clouds and galactic systems remains obscure in many respects. In the foregoing sections we have pointed out contrasts and similarities, alluding to the evidence for a massive galactic nucleus, the singularities in the distribution of galactic clusters, and the high-speed motion of the local system with respect to the globular clusters.

The earlier hypothesis (Section 68) concerning the possible origin and growth of the Galaxy has not been disproved. Nevertheless, I think there is increasing evidence from outside the galactic system that the interpretation should be amended. There were difficulties with the earlier theory that are now avoided or resolved—such, for instance, as its incompleteness in leaving the spiral nebulae out of the picture, the scarcity of clusters intermediate between globular and galactic groups, the singular distribution of dark nebulosities, and the slowness or even impossibility of amalgamating clusters and star clouds with no more potent resisting medium available than the galactic star field with its infrequent encounters.

In brief, I propose the following picture of the galactic system,[20] realizing, of course, that our researches in the next few years, guided in part by the present hypothesis, may modify or remake the picture.

1. Our galactic system, it now appears, is neither a spiral, such as the Andromeda Nebula (Messier 31), nor a single unified discoidal star system, like a Magellanic Cloud on a grand scale;

[20] H. Repr. 61, 1929.

it is rather a super-galaxy—a flattened system of typical galaxies.

2. In mass and population, therefore, the galactic system should be compared with the Coma-Virgo Cloud of bright galaxies, rather than with one of its members. Our local system, a star cloud that is a few thousand light years in diameter, appears to be a galaxy, similar to the Clouds of Magellan or to a typical extra-galactic nebula. That all types of spirals and Magellanic Clouds are systematically smaller than our galactic system seems clearly established by recent work on star clouds and on the Coma-Virgo system of nebulae.[21]   Five to ten thousand light years appears to be the average diameter, and this is not over four or five per cent of the diameter of the galactic system.

3. The Scutum star cloud, the Cygnus star cloud, and a half dozen or so other distinct Milky Way star clouds are, or have been, on this interpretation, typical galaxies, in the sense in which the average spiral nebula is called a galaxy.

4. The three or four clouds of galaxies in Coma-Virgo appear to have diameters approximating 2,000,000 light years, but many of the less populous systems have diameters well under 1,000,000 light years; such, for instance, are the Ursa Major group measured by Baade,[22] and the Pegasus group, N. G. C. 7317 to 7320.

5. In comparing our galactic system with a cloud of galaxies, we note that it is considerably flattened and seems to be unusually compact. But a newly discovered cloud of four or five hundred galaxies in Centaurus likewise has a projected length three or four times its width;[23] many of its component systems are apparently in contact—a phenomenon that is occasionally observed in the Coma-Virgo clouds and elsewhere.

6. If our galactic system is a flattened cloud of galaxies, we may raise the question as to the continuity of galactic star fields

[21] Shapley and Ames, H. C. 294, 1926; H. B. 864, 865, 866, 868, 869, 1929; H. B. 873, 1930.

[22] Hamb. Mitt., 6, No. 29, 103, 1928.

[23] Shapley, H B. 874, 1930.

between the separate galaxies and the meaning of the galactic nucleus that is marked out by globular clusters, novae, star clouds, and the distribution of faint stars. It may be noted that Pannekoek and others have pointed to anomalies in star distributions, which, I think, may in part be accounted for by large breaks in the sheet of galactic star clouds; but these irregularities in star distribution are partly features of the local system—we are perhaps beginning the detection of the present structure of our own spiral.

7. Viewing our local system as a galaxy—possibly at one time a typical ellipsoid or spiral—we observe that (a) the major part of the recorded obscuring nebulosity is concentrated in the plane of the local system; also, that nearly all the conspicuous dark regions appear to be within 1,000 parsecs or so of the sun, a distance which indicates that they are a remnant of the peripheral dark ring of matter such as is observed in many spiral nebulae;[24] (b) the inclination of the local system indicates the original equatorial plane of the ellipsoid or spiral; (c) when observed from the distance of the Coma-Virgo galaxies the Orion Nebula with its involved stars, the Pleiades, and other groups near the sun would appear as nebulous knots in the structure of our local system; (d) from an outside point the local system and the Cygnus star cloud would appear as galaxies in collision.[25]

8. The great star cloud in the Sagittarius region has interesting analogies with the Andromeda Nebula. Its dimensions appear to be approximately the same; novae are frequent, but long-period Cepheids are scarce near its center. If Kepler's nova (Nova Ophiuchi No. 1, $17^h 24^m.6 - 21° 24'$ (1900)) is at the distance of the cloud its absolute magnitude at maximum was about the same as that of S Andromedae—the abnormal nova of 1885. At the distance of the Andromeda Nebula the angular separations of stars in the Sagittarius Cloud would be only six per cent of those now observed, the stars would be six magni-

[24] Curtis, Lick, Publ., 13, 9, 1918.
[25] Shapley, H. Repr. 8, 1924.

tudes fainter, and the densest central region would be unresolved with present telescopic power.

In brief, it is proposed that our galactic system is a cloud of ordinary galaxies, the considerable oblateness of the cloud representing a stage in its normal collisional development. This hypothesis makes our Galaxy less anomalous than it appears on previous views, which conceived ours either as an enormous discoidal star cloud unduplicated by any other visible organization or as a spiral nebula forty or fifty times the diameter of the average spiral system.

CHAPTER XV

A PARTIAL SUMMARY

Much of our present information on the dimensions of sidereal systems, the transparency of space, the relative frequency of supergiant stars, and the distance and direction of the galactic center has developed from the study of star clusters, which have also contributed effectively to research on such subjects as Cepheid variation and the velocity of light. A brief synopsis of the more interesting and significant results attained in the course of the work described in the preceding chapters is given below. The summary may be conveniently made under the heads Variable Stars, Clusters, and the Galaxy.

Variable Stars

In a study, not yet complete, of 45 globular clusters (43.7 per cent of all now known) 886 variables have been found, mainly by investigators at Harvard and Mount Wilson (see Table IV, I). An unaccountable inequality is found in the number of variable stars; 6 per cent of the clusters are wholly barren of variables; 41 per cent contain no more than five each; whereas in others, of the same neighborhood and class, scores of variables are known—in some more than 10 per cent of all the giant stars show variation. Variability is unknown among the low-luminosity stars in clusters, but the tests have been only preliminary, except in ω Centauri, Messier 13, and one or two others.

Of the 886 known variables, about 50 per cent have been shown to belong to the cluster type (that is, they are Cepheids with periods less than a day); many of the remainder probably are also of the same type, but their light curves have not been measured or the periods found. These short-period Cepheids

are widely scattered, some appearing quite beyond the normally
recognized limits of the clusters.   The lack of a high concentration to the center may be a result of high velocities; but practically nothing is yet known observationally of the differential
motions in globular clusters.   (See papers by van Maanen and
Balanowsky in Appendix C.)

There are a number of classical Cepheids in globular clusters,
a few long-period variables, and a few irregular variables, but
as yet no certainly verified eclipsing stars.   The frequency of
cluster-type Cepheids relative to that of classical Cepheids is
greater in clusters than near the sun, but various factors of
observational selection disturb the comparison.   Investigations
show that the light curves, ranges of period, and color phenomena for Cepheid variables in clusters and in the solar neighborhood are essentially the same, although there are unexplained
differences in relative frequency of different types due, possibly,
to environment or age.

Few variable stars are associated with galactic clusters,
perhaps none are actual members.   The relatively strong
galactic concentration of the classical Cepheids of the galactic
system is well known.   The absence of galactic concentration
for cluster-type variables may be attributable to the high velocities.   (The space velocities of classical Cepheids are normally
low.)   Likewise, the large proper motions of many of the
cluster-type Cepheids of the galactic system should no longer
be taken as indicating nearness and low luminosity, since their
radial velocities prove their rapid motion in space (Table X,
XIII). ´

The relation of logarithm of period to absolute magnitude,
called the period-luminosity curve, has been extended, from
the preliminary work of Miss Leavitt on apparent magnitudes for 25 variable stars in the Small Magellanic Cloud, to
several hundred stars in a dozen star clusters and clouds
(Chapter X).   It is found by Hubble to hold also in some extra-
galactic nebulae.   The great majority of Cepheids fit in with

this empirical period-luminosity curve; the exceptions are few but probably of much importance. The curve can be calculated rather satisfactorily on the basis of the pulsation theory of Cepheid variation, though the latter is hardly complete. The visual period-luminosity curve, developed and used in my earlier discussion of the distances of clusters, is now fully replaced by the photographic period-luminosity curve derived from the Magellanic Clouds and globular clusters.

With the period-luminosity curve, the relative distances of all normal Cepheid variables can now be determined with considerable accuracy from measures of periods and apparent magnitudes. The absolute values of the distances will become equally accurate when current investigations of the proper motions and radial velocities of the nearer Cepheids have fixed securely the zero point of the period-luminosity curve.

An important relation has also been found between length of period and the spectral class or color—the period-spectrum curve—which holds generally for classical galactic Cepheids. The longer the period, the redder and brighter is the variable—a clear connection of low density with high luminosity for giant stars (Section 49). In one direction the period-spectrum relation extends to the cluster-type Cepheids, and in the other through RV Tauri variables to the typical long-period variables. On the basis of the period-spectrum curve for galactic Cepheids, a relation between absolute magnitude and period can be calculated (Section 51) which is identical with the observed period-luminosity relation and shows again that galactic Cepheids and those of clusters, Magellanic Clouds, and extragalactic nebulae are thoroughly comparable.

To test the accuracy of the dimensions of the Galaxy as derived from globular clusters, an extensive investigation of the faint variable stars in the Milky Way is in progress at the Harvard Observatory. The study has resulted in the discovery of two or three thousand new variable stars, in the determination of the distance to the star clouds in Sagittarius which appear to compose the galactic nucleus, and in the acquisition of much material for statistical and descriptive studies of variable stars of all types. It is found that the long-period variable star, which is a giant at maximum, may be of great importance in future investigations of the distances of the Milky Way star clouds and of such star clusters as contain variables of this class (Section 69).

The cluster-type variables in distant systems afford a means of testing with extraordinary accuracy the relative velocity of blue and yellow light. No difference in speed is found throughout a journey of 40,000 years for light waves differing in length by 25 per cent.

A difficulty for all Cepheid theories that involve the gravitational relation between period and mean density $(P^2 \propto 1/\bar{\rho})$ is found in the essential constancy of median magnitude, color, and surface brightness for all periods less than one day (Sections 21 and 22). Probably nuclear differences are concerned in the production of the variety of periods. There appears to be a possibility, through careful photometry of globular clusters, of witnessing the beginning or the dying out of Cepheid variation in stars that are much like these cluster variables in color and magnitude (Section 23).

CLUSTERS

Whatever its meaning and interpretation, the period-luminosity curve for Cepheids affords a powerful practical method of measuring the distances of all stellar systems— globular clusters, star clouds, and nebulae—that contain Cepheid variables. It has been possible to develop other useful methods for measuring the distances of clusters, employing variously the magnitudes, spectra, and colors of stars of high luminosity, as well as integrated apparent magnitudes and angular diameters. By using two or more of these various methods the distance of the average globular cluster is determined with an estimated probable error of 15 per cent (aside from uncertainty of the zero point).

Many contributions to knowledge of the transparency of space have been made in the course of the study of clusters. The obstruction of light by dark nebulae in special regions is well known. The differential scattering of light in interstellar space is found to be negligible in both high and low galactic latitudes (Chapter IX). General obstruction (non-selective diminution of light) throughout space is neither proved nor disproved, but it appears on present evidence to be of small practical importance in the measurement of distances up to 100,000,000 light years.

The distances of the globular clusters now known range from 18,000 to 184,000 light years. The galactic clusters are, on the average, much nearer; but for many of them the distances are not yet accurately known. Ninety-three globular systems are catalogued in Appendix A. Ten others are members of the Magellanic Clouds (Table XIII, I). The globular clusters appear very definitely to be a part of the Galaxy. They are distributed throughout an oblate system, symmetrical with respect to the galactic plane. Although they are concentrated toward the galactic plane, only two are actually within four degrees of it. The galactic clusters, on the other hand, are almost exclusively in the low latitudes that are essentially devoid of globular clusters.

The globular clusters seem to be fairly uniform in size, content, and total intrinsic brightness. There is diversity in central condensation, however, and occasionally other divergences from normal structure appear. On the basis of these variations, a scheme of 12 classes of globular clusters has been developed (Section 5).

The diameters of typical globular clusters are of the order of 35 parsecs (Chapter XIV). The galactic clusters are, on the average, much smaller, but they are more varied in size and richness. An increase in diameter with decreasing brightness is shown for galactic clusters, giving rise in clusters such as Messier 67 and the Pleiades to the appearance of a concentrated nucleus in a widely dispersed larger system (Section 34). The same wide dispersion of the fainter and less massive stars apparently holds for globular clusters.

A catalogue of 249 galactic clusters is given in Appendix B, with new data on magnitudes, dimensions, orientation, and distances (see Sections 7 and 57). In making this catalogue, a new and comprehensive classification of galactic clusters based on richness and apparent concentration has been developed and employed.

The maximum luminosities of stars in globular clusters, if the distances which I derive are not too small, rarely if ever exceed −3.5 visually, and therefore do not attain the high values found in the Magellanic Clouds and in the galactic clusters of B stars. If we except a few stars, the maximum is, in fact, about −2.5. The classical Cepheids are always found among the brightest objects; in the clusters and in the Magellanic Clouds some are of extraordinary luminosity, many of them exceeding absolute visual magnitude −4.

For several globular clusters a preliminary maximum in the general luminosity curve is found at or near zero absolute magnitude. There may be some association of the observed excess (which is probably composed nearly altogether of white stars) with the cluster-type variables (Section 27) and with the fainter part of the period-luminosity curve. But to interpret this phenomenon fully we need further detailed work on magnitudes and colors.

The globular clusters in the Large Magellanic Cloud show a spread in integrated absolute photographic magnitude from −6.5 to −9.1, but if two of the objects, which possibly are not globular cluster members of the Cloud, are excluded, the extreme values are −6.5 and −7.2. The only cluster in the Small Magellanic Cloud that appears certainly globular has the absolute photographic magnitude −7.1 (Section 66). The assumption of the general comparability of globular clusters is supported by this relatively small dispersion, though "giant-poor" globular clusters must be segregated in using magnitudes and angular diameters for the determinations of relative distances (Sections 55 and 56).

There is a scarcity of forms intermediate between globular and galactic clusters, but with further research it should be possible to work out the details of the relationship between the types and an explanation of the peculiar contrast in space distribution. Obscuring nebulosity throughout the Galaxy will play a part in this explanation.

From star counts and from small-scale photographs it is found that many globular clusters are distinctly elliptical in projected outline, probably being oblate spheroids. The data in Chapter VI indicate, indeed, that most of the globular clusters are non-spherical, and for 37 the degree and orientation of the flattening is definitely measurable on average photographs.

There are tempting analogies between the distribution of stars in globular clusters and certain features in the kinetic theory of a gas; but the observations are regrettably insufficient and the approach to such a comparison must be cautious. In the attempt to find density laws we encounter three seriously perturbing factors, discussed in Chapter V: (a) non-sphericity of clusters and the impossibility of evaluating the true ellipticity, with the consequence, therefore, that the differences in the density laws along the various radii must be ignored; (b) the Eberhard effect in condensed regions; (c) the magnitude limitations, which make us attempt interpretations of the distribution of a million or so dwarfs and giants on the basis of partial counts of a few hundred, or a thousand or so, high-luminosity (massive) stars. But whatever the true density laws, there appears to be a preferential concentration of bright stars (possibly massive objects) at the centers of globular clusters.

Stars of all common colors and spectral classes have been recorded in both globular and galactic clusters, and in the latter various degrees of nebulosity are found. Many kinds of peculiar stars are also present. It seems that in clusters practically all the materials exist that constitute star clouds and galaxies.

In spectral composition the galactic clusters of the Milky Way are of two principal kinds (Trumpler proposes a more elaborate subdividing): (a) the Hyades model, with giant red-yellow stars as well as white giants and the main series of ordinary dwarf stars; (b) the Pleiades model, with no yellow-red giants or supergiants, all the stars falling along the main branch of a Russell diagram, the redness increasing with decreasing brightness. For both kinds of cluster, supergiant stars are relatively uncommon, occurring most frequently as class B stars in the Pleiades type.

In all globular clusters so far investigated the most luminous stars are found to be red, with a progressive change toward blueness for the next few fainter magnitudes. This relation of magnitude to color appears to differ from the average conditions for galactic giant stars; but in the Galaxy the general non-homogeneity is obvious and the stellar sources are much confused. From the color-magnitude arrays (Section 12) it is possible, on fairly reasonable assumptions, to compute the mean masses of the giant stars in globular clusters for various intervals of color or spectrum. The computation of average masses for Messier 22 (Section 73) shows values in terms of the sun's mass as follows: M0, 24.0; K0, 10.8; G0, 4.9; F0, <3.6; A0, <4.8.

The coexistence in the globular clusters of blue stars fainter than absolute magnitude zero and red supergiants brighter than −2.5, and also the yellow and white giant stars in the Hyades and other galactic clusters, gives rise to inquiries concerning the speed and direction, or even the generality, of the evolution of stars. The current theories of giant stars require that the duration of the total giant stage, if contraction is the main source of the energy of radiation, can scarcely exceed 100,000 years. But the color-luminosity relation appears to be identical in near and remote clusters, though because of the finite velocity of light many such clusters differ from each other in age by more than 100,000 years. There is thus little if any evidence of ageing in the typical globular clusters; stellar evolution is almost immeasurably slow, and contraction, of course, must be abandoned as the source of the energy of radiation. The somewhat anomalous consequences are discussed in Section 74.

THE GALAXY

Minimum dimensions of the galactic system are found from the distances separating some of its stars and star clusters. Revising earlier values, we now find 70 kiloparsecs as the diameter in the galactic plane (Chapter XII). The diameter perpendicular to the plane is probably about one tenth as great, with, however, a quantity of far outlying stars that are probably affiliated (Section 62).

The center of the Galaxy, judged from the distribution of clusters and of several kinds of high-luminosity stars, is in the general direction of 17^h 30^m, $-30°$ (Sagittarius) and approximately 16 kiloparsecs distant from the sun. Around this massive nucleus (Section 61) the neighboring galactic stars appear to move with a velocity of some 300 kilometers a second and a period well in excess of 100,000,000 years.

Examination of the colors in the distant Milky Way star clouds reveals the presence of stars of all common spectral types, indicating that these remote regions probably do not differ fundamentally from the solar neighborhood.

The brighter early class stars—the B's, and many A's—are members of a local cloud, which contains also many stars of other types and some of the nearby galactic clusters and diffuse nebulae. The local system is also considerably flattened, and its central plane is found, in confirmation of Gould's work on the bright stars and Charlier's investigation of the brighter B's, to be inclined some 10 or 15° to the Milky Way plane. The sun, as Newcomb and others have shown, is north of the galactic plane (hence the dip of the galactic circle); it is also about 55 parsecs above the central plane of the local cloud and 90 parsecs from the center which lies in the direction of Carina, approximately at right angles to the direction to the center of the Galaxy. Other values of the distance of the sun to the central planes and to the center of the local system have been derived by different investigators; the most dependable work gives distances and directions in general agreement with those quoted above.

In variety of stars, nebulae, and clusters, and in irregularity of structure, the Magellanic Clouds afford a fair parallel to our own Galaxy; in dimensions, however, they probably approach the star clouds of the Milky Way, and of course are much inferior to the galactic system in diameter, volume, and total mass. They are of the same general diameter as many external galaxies (extra-galactic nebulae), and though differing from true spirals and elliptical nebulae in form, they appear to be exactly comparable with a small subclass of irregular extra-galactic nebulae.

A higher organization is shown in the Coma-Virgo cloud of external galaxies, which contains 300 or more true spirals, double nebulae, spherical nebulae, and various sorts of elliptical and spindle systems. The distance of this supergroup is found to be of the order of 3,000,000 parsecs. A number of more remote clouds of extra-galactic nebulae are on record, some at distances in excess of 40,000,000 parsecs, but they are not yet investigated in detail.

In total mass and stellar population, the galactic system is more nearly comparable to the whole Coma-Virgo cloud than to any one of its individual members. In the last section of the preceding chapter the tentative theory is advanced that our galactic system is in fact a super-galaxy—a flattened cloud of ordinary galaxies.

APPENDIX A

CATALOGUE OF GLOBULAR CLUSTERS

The material on which is based the catalogue of globular clusters has been described in the text. For three clusters, N. G. C. 4372, N. G. C. 6356, and N. G. C. 6864, special notes appear at the end of the catalogue. Daggers after the N. G. C. numbers in the first column indicate questionable objects (Section 6). Galactic coordinates are on the Harvard system (pole $12^h 40^m$, $+28°$). The values of the ellipticity and orientation are described in Section 30. The magnitudes and diameters are explained in Chapter XI, and in that chapter and in Chapter XII is described the derivation of the distances listed in the last three columns of the catalogue.

The distance of N. G. C. 4147 is 19.5 kiloparsecs according to a preliminary study of its two cluster-type Cepheids reported in a letter from Dr. Baade.

CATALOGUE OF GLOBULAR CLUSTERS

N. G. C.	Name	R A 1900	Dec 1900	Galactic		Angular Diameter	Integrated Magnitude	Class	Spectrum	Number of Variables
				Long	Lat					
		h m	° '	°	°	'				
104	47 Tuc	0 19 6	−72 38	272	−45	23	3	III	G5	7
288		0 47 8	−27 8	157	−88	10 0	7 2	X		2
362	Δ 62	0 58 9	−71 23	268	−47	5 3	6 0	III	G5	14
1261		3 9 5	−55 36	237	−51 5	2 0	8 5	II	G	
1851	Δ 508	5 10 8	−40 9	212	−34 5	5 3	6 0	II		3
1904	M 79	5 20 1	−24 37	194	−28	3 2	8 1	V		5
2298	. .	6 45 4	−35 54	213	−15	1.8	10 1	VI		
2419	...	7 31 4	+39 6	148	+23	1 7	11 0	VII		.
2808	.....	9 10 0	−64 27	249	−11	6 3	5 7	I	Ko	
3201	Δ 445	10 13 5	−45 54	244	+ 9	7 7	7 4	X		61
4147	. ..	12 5 0	+19 6	226	+79	1 7	10 3	IX	A7.	5
4372ⁿ	......	12 20 1	−72 7	269	−10	12 0	7 8	XII		
4590	M 68	12 34 2	−26 12	269	+36	2 9	7 6	X		28
4833	. ...	12 52 7	−70 20	271	− 8 5	4 7	6 8	VIII		5
5024	M 53	13 8 0	+18 42	305	+79	3 3	6 9	V		40
5053	. .	13 11 5	+18 13	309.5	+78	3 5	10 5	XI		9
5139	ω Cen	13 20 8	−46 47	277	+15	23	3	VIII		132
5272	M 3	13 37 6	+28 53	8	+77 5	9 8	4 5	VI	G	166
5286	Δ 388	13 39 9	−50 52	280	+10	1 6	8 5	V	Go	0
5466	.	14 1 0	+29 0	8	+72 5	5 0	10 0	XII		14
5634	......	14 24 4	− 5 32	310 5	+48 5	1 3	10 4	IV		
4499	......	14 45 0	−81 49	275	−20	3 1	11 5	XI		
5824	...	14 57 8	−32 40	301	+21	1 0	9 3	I	F8	
5897	....	15 11 7	−20 39	312	+29	7 3	7 7	XI		
5904	M 5	15 13 5	+ 2 27	332	+46	12 7	3 6	V	G	84
5927	. ..	15 20 8	−50 19	294	+ 4	3 0	8 8	VIII		
5946†	...	15 28 2	−50 19	295 5	+ 3.5	1 3	10 6	IX.		
5986	Δ 552	15 39 5	−37 27	305	+12	3 7	7 0	VII	F8	1
6093	M 80	16 11 1	−22 44	320.5	+18	3 3	6 8	II	Ko	4
6101	...	16 14 4	−71 58	284 5	−16	3 8	9 5	X		
6121	M 4	16 17 5	−26 17	319	+15	14 0	5 2	IX	F	33
6139	.....	16 21 0	−38 36	310	+ 6	1 3	9 8	II		
6144	..... .	16 21 1	−25 49	319	+15	3 3	10 3	XI		
6171	..	16 26 9	−12 50	332	+22	2 2	8 9	X		
6205	M 13	16 38 1	+36 39	27	+40	10 0	4 0	V	Go	7
6218	M 12	16 42 0	− 1 46	344	+25	9 3	6 0	IX		
6229	......	16 44 2	+47 42	40	+40	1 2	9 7	VII·	.	1
6235	. ..	16 47 4	−22 0	327	+12	1 9	10 8	X		
6254	M 10	16 51 9	− 3 57	343	+22	8 2	5 4	VII		
6266	M 62	16 54.9	−29 58	322	+ 7	4 3	7 0	IV	Ko	26
6273	M 19	16 56 4	−26 7	324 5	+ 9	4 3	6 8	VIII	G5	.
6284	.....	16 58 4	−24 37	326	+ 9	1 5	10 0	IX	F.	
6287	. ..	16 59 1	−22 34	328	+10	1 7	10 4	VII		
6293	.....	17 4 0	−26 26	325	+ 7	1 9	8 8	IV	G5	3
6304	. . .	17 8 2	−29 20	323	+ 5	1 6	9 2	VI	K:	
6316		17 10 3	−28 1	325.5	+ 5	1 1	9.9	III	G5	
6325	......	17 11 9	−23 38	327 5	+ 6	0 7	11 9	IV:	.	
6333	M 9	17 13 3	−18 25	333	+10	2 4	7 4	VIII	K?	1
6341	M 92	17 14 1	+43 15	36	+35	8 3	5 1	IV	G5:	14

CATALOGUE OF GLOBULAR CLUSTERS —(continued)

N. G. C.	Ellipticity	Orientation	Photographic Magnitude				Adopted Modulus	Quality	Distance	$R \sin \beta$	$R \cos \beta$
		°	Var.	Bright	6th	30th			kpc	kpc	kpc
104	8	−55		13 09	12 4	13 4	14 17	b	6 8	− 4 8	4 8
288	9			14 80	14 5	15 1	15 81	b	14 5	−14 5	0 5
362	8	+65	15 5	14 12	13 5	14 8	15 55	b	12 9	− 9 4	8 8
1261	9 5						16 72	c	22 0	−17 2	13 7
1851	9	−75					15 78	c	14 3	− 8 1	11 7
1904	9	+ 5		15 29	15 01	15 72	16 54	b	20 4	− 9 6	18 0
2298	8	+39					17 12	d	26 5	− 6 9	25 6
2419	9	−56					17 41	d	30 3	+11 9	28 0
2808	8	+84		14 9	14 3	15 4	16 05	b	16 3	− 3 1	16 0
3201	9		14 52	13 52	13 3	13 8	14 81	c	9 2	+ 1 4	9 1
4147			16 8	16 58	16 23	16 93	16 93	b	24 2	+23 7	4 6
4372	9						14 91	e	9 6	− 1 7	9 5
4590	9		15 90	14 80	14 31	15 08	15 95	a	15 5	+ 9 1	12 6
4833	8	−80	.	....			16 01	c	15 9	− 2 2	15 7
5024	9	−79*		15 07	14 94	15 26	16 30	a	18 2	+17 9	3 5
5053	8	−61	16.19	15 65	15 1	16 0	16 20	a	17 3	+17 0	3 6
5139	8	+30	14 37	12 91	12 6	13 1	14 15	b	6 8	+ 1 8	6 6
5272	8	+54	15 50	14 23	13 92	14 45	15 43	a	12 2	+11 9	2 6
5286	9 5						16 80	d	23 9	+ 4 2	23 5
5466	9		16 17	15 72	15 1	16 2	16 16	b	17 0	+16 2	5 1
5634	9						17 49	c	31 4	+23 2	20 5
I 4499	9			.		....	16 91	e	24 1	− 8 2	22 7
5824	.			. .			17 32	e	29 1	+10 4	27 2
5897	8	−44		15 15	14 9	15 4	16 07	b	16 4	+ 8 0	15 5
5904	9	+16	15 26	13 97	13 74	14 27	15 17	a	10 8	+ 7 8	7 5
5927	9						16 56	d	20 5	+ 1 4	20 1
5946†	9					.	17 54	d	32 2	+ 2 0	32 1
5986							16 14	c	16 9	+ 3 5	16 6
6093	10	. .		14 88	14 72	15 09	16 22	a	17 5	+ 5 4	16 6
6101	8	+35					16 60	d	20 8	− 5 7	20 0
6121	9	+72*	14 27	13 88	13 3	14 4	14 30	b	7 2	+ 1 9	7 0
6139	9	−64					17 34	d	29 3	− 3 1	29 1
6144	8	−22		15 76	15 2	16 3	16 29	b	18 1	+ 4 8	17 5
6171	9	..		15 46	15 2	15 9	16 63	b	21 2	+ 7 9	19 6
6205	9.5	−63*	15 20	13 75	13 45	13 92	15 07	a	10 3	+ 6 6	7 9
6218				13 97	13 56	14 31	15 21	b	11 0	+ 4 6	9 9
6229	.	..		16 18	15 90	16 37	17 37	b	29 8	+19 2	22 8
6235	8	+89		16 17	15 7	16 8	17 28	c	28 6	+ 5 9	28 0
6254	9			14 06	13 35	14 38	15 26	b	11 2	+ 4 2	10 4
6266	8	+16	16 40	15 87	15 6	16 1	16 35	c	18 6	+ 2 3	18 4
6273	6	−28					16 06	c	16 3	+ 2 6	16 1
6284	10						17 24	c	28 0	+ 4 4	27 7
6287	9	.	.				17 24	d	28 0	+ 4 9	27 5
6293	9						16 82	c	23 1	+ 2 8	22 9
6304	9 5	.					17 02	c	25 3	+ 2 2	25 2
6316	9						17 52	d	31 8	+ 2.8	31 6
6325	.	.		...	....	.	18 3	e	46	+ 4 8	46·
6333	9			15 61	16.76	15 08	16 61	b	20 8	+ 3 6	20 5
6341	8	+16.		13 86	13 60	14 16	15 24	b	11 2	+ 6 4	9 2

CATALOGUE OF GLOBULAR CLUSTERS —(continuea)

N G.C	Name	R A. 1900	Dec 1900	Galactic Long	Lat.	Angular Diameter	Integrated Magnitude	Class	Spectrum	Number of Variables
		h m	° '	°	°	'				
6342	.. ..	17 15 3	− 19 29	333	+ 8	0.5	11.4	IV	..	
6352†	..	17 17 5	− 48 22	309	− 8	2 5	7.9	XI·	..	..
6356ᵃ	.	17 17 8	− 17 43	334	+ 9	1 7	8.6	II	Ko	
6362	Δ 225	17 21 5	− 66 58	293	− 18	6 7	7 1	X	..	17
6366	...	17 22 4	− 4 59	346	+15	4:	12.1	XI	..	..
6388		17 29 0	− 44 40	313	− 8	3 4	7.1	III	K	.
6402	M 14	17 32 4	− 3 11	349	+14	3 0	7 4	VIII	...	..
6397	Δ 366	17 32 7	− 53 37	304 5	−12 5	19 0	4 7	IX	G?	2
6426†	.....	17 39 9	+ 3 13	356	+15	1 3	12 2	IX	..	..
6440	....	17 43 0	− 20 20	335	+ 2	0 7	11 4	V	..	..
6441	.. ...	17 43 4	− 37 1	321	− 6 5	2.3	8 4	III	Ko	.
6453	..	17 44 7	− 34 36	322.5	− 5 5	0 7	11 2	IV	...	
6496	.	17 51 8	− 44 14	315	−10 5	2 2	9 7	XII	...	
6517	.. . .	17 56 4	− 8 57	347	+ 6	0 4	12 1	IV	...	..
6522	.	17 57 2	− 30 2	328	− 5	0 7	11.0	VI	...	
6528	.. .	17 58 4	− 30 4	328 5	− 5	0 5	11 8	V	...	.
6535†	. ..	17 58 7	− 0 18	355	+ 9 5	1 3	11 9	XI.	...	..
6539†	.	17 59 4	− 7 35	348	+ 5	1 3	12 6	X:	..	1
6541	Δ 473	18 0 8	− 43 44	317	−12	6 3	5 8	III	G	1
6553	.	18 3 2	− 25 56	332 5	− 5	1 7	10 0	XI	...	0
6569	.	18 7 2	− 31 51	328	− 7 5	1 4	10 2	VIII	..	
6584	Δ 376	18 10 6	− 52 15	309 5	−17	2 5	8 3	VIII	..	0
6624	.	18 17 3	− 30 24	330	− 10	2 0	8 6	VI	Mo	
6626	M 28	18 18 4	− 24 55	335	− 7	4 7	6 8	IV	G5	9
6637	M 69	18 24 8	− 32 25	329	− 11	2 8	7 5	V	K2	
6638	..	18 24 8	− 25 34	335 5	− 7.5	1 4	9 2	VI		
6652	.	18 29 2	− 33 4	328 5	− 13	1 7	8 7	VI:	K5	.
6656	M 22	18 30 3	− 24 0	337	− 9	17 3	3 6	VII	.	21
6681	M 70	18 36.7	− 32 23	330	− 13 5	2.5	7.5	V	..	..
6712†	. .	18 47.6	− 8 50	353 5	− 6	2 1	9 9	IX.	.	1
6715	M 54	18 48.7	− 30 36	333	− 15	2 1	7 1	III	F8	.
6723	Δ 573	18 52 8	− 36 46	328	− 19	5.8	6.0	VII	G5?	17
6752	Δ 295	19 2 0	− 60 8	303	− 26 5	13 3	4 6	VI	Go	1
6760†	.	19 6 1	+ 0 52	3	− 5	1 9	10 9	IX	. .	.
6779	M 56	19 12 7	+30 0	30	+ 8	1 8	8 8	X	. .	1
6809	M 55	19 33 7	−31 10	336	−25	10.0	4 4	XI	. .	2
6864ᵃ	M 75	20 0 2	−22 12	347	−27	1 9	8 6	I	Go	11
6934	.	20 29 3	+ 7 4	20	−20	1 5	9 4	VIII	Go	
6981	M 72	20 48 0	−12 55	3	−34	2 0	8 6	IX	...	29
7006		20 56 8	+15 48	32	−21	1 1	11 8	I	..	11
7078	M 15	21 25 2	+11 44	33	−28	7 4	5 2	IV	F	74
7089	M 2	21 28 3	− 1 16	21	−36	8 2	5 0	II	F5	10
7099	M 30	21 34 7	−23 38	355	−48 5	5 7	6 4	V	F8	3
7492	.	23 3 1	− 16 10	22	−64	3 3	10 8	XII	..	9

NOTES TO APPENDIX A

N. G. C. 4372. The distance depends only on the diameter measure. The cluster appears to be partially obscured by one of the streamers from the Coal Sack nebula. A special investigation of the magnitudes is being made on Harvard plates

N. G. C. 6356, 6864. The distance depends only on the magnitudes of the bright stars, since the integrated brightness and diameter appear to be abnormally large There may be obscuring nebulosity in the field of N G. C 6356

CATALOGUE OF GLOBULAR CLUSTERS.—(*continued*)

N.G.C	Ellip-ticity	Orien-tation	Photographic Magnitude				Adopted Modulus	Qual-ity	Dis-tance	$R \sin \beta$	$R \cos \beta$
			Var	Bright	6th	30th					
		°							kpc	kpc	kpc
6342							18.0	e	40.	+ 5 6	39 5·
6352†	9						16 48	d	19 7	− 2 7	19 4
6356	9	− 14		17 16	16 86	17 44	18 51	e	50	+ 7.8	49.
6362	8	+ 78					15 90	d	15 1	− 4 7	14.4
6366							17 34	e	29.	+ 7 5·	28.
6388	9 5						16 20	c	17 3	− 2 4	17 2
6402	9	+ 76		15 44	14 85	15 86	16 48	a	19 7	+ 4 8	19 1
6397	9	+ 73		12 61	11 9	13 1	13.76	c	5 65	− 1.2	5 5
6426†	9						17 85	e	37 1	+ 9 6	35 9
6440	8	+ 10					18 5	e	50:	+ 1 7.	50·
6441	8	+ 40					16 62	c	21 1	− 2 2	21 0
6453							18 5	e	50.	− 4 8.	50:
6496							16 90	d	24 0	− 4 0	21 6
6517	8	− 4					18 5·	e	50	+ 5 2	50:
6522							17 78	e	36 0	− 3 2	35 8
6528							18 24	e	44 4	− 3 9	44 1
6535†				15 9	15 3	16 4	17 13	d	26 7	+ 4 4	26 3
6539†	9						17 94	e	38 7	+ 3 4	38 4
6541	9		14 47	13 35	12 7	13 8	14 76	c	8 9	− 1 8	8 7
6553	9						17 14	c	26 8	− 2 3	26 7
6569	9 5			.			17 35	c	29 5	− 3 8	29 2
6584	9		·	. .	. .		16 56	c	20 5	− 6 0	19 6
6624	10		·	. .	. . . .	·	16 74	c	22 2	− 3 8	21 8
6626	9	+ 18	. . .	14 87	14 49	15 11	16 10	a	16 6	− 2 0	16 4
6637	9		. .	. .	. . . . . .	. .	16 36	c	18 7	− 3 6	18 4
6638	8	− 27	. .	16 22	15 90	16.60	17 36	b	29 6	− 3 9	29 3
6652	8	− 11	·	.	. . .	. . . .	16 87	d	23 6	− 5 3	23 0
6656	8	+ 18	14 06	12 93	12.80	13 26	14 16	a	6 8	− 1 1	6.7
6681	9 5			.	. . .	. . .	16 41	c	19 2	− 4.5	18 8
6712†				16 10	15 65	16 36	17 10	d	26.2	− 2.7	26.0
6715	10		. .		. .	·	16 44	d	19 4	− 5 0	18 7
6723	9 5		15 33	14 20	13.7	14 8	15 44	b	12 3	− 4 1	11.7
6752			. . .	13 26	12 8	13.6	14 62	b	8 4	− 3 8	7 4
6760†			. . .	·	. .	. . . .	17 28	e	28 6	− 2 5	28 5
6779	8	+ 12	. .	15 31	14 98	15 70	16 54	b	20 3	+ 2 8	20 0
6809	9		. .	13 58	12 9	14 2	14 74	b	8 8	− 3 7	8 0
6864	9		. .	17 06	16 76	17 35	18 43	d	48 5	− 22 0	43 2
6934	9		.	15 78	15 33	16 11	16 98	b	24 9	− 8 5	23 4
6981			.	16.80	15 86	15 53	16.20	16 84	a	23 3	− 13 0	19 3
7006				18.96	17 50	16 99	17.89	18 77	b	56 8	− 20 4	53 0
7078	8	− 11	15.63	14.31	14 13	14.55	15 60	a	13 1	− 6.1	11 6
7089	9	− 80*	15.71	14 61	14 25	14 76	15 72	a	13 9	− 8 2	11 3
7099	9		. . .	14 63	13 77	15 04	15 82	b	14 6	− 10 9	9.7
7492	9		.	16 82	16 3	17 1	17 01	c	25.2	− 22 7	11 1

APPENDIX B

CATALOGUE OF GALACTIC CLUSTERS*

N G C	R.A. 1900 (h m)	Dec. 1900 (° ')	Galactic Long.	Galactic Lat	Class	Approx. Diameter (')	Orientation (°)	Number of Stars	Mag. of Lim. of Plate	Mag. of 5th Star	Distance M=+0.5 (kpc)	Distance M=-0.5 (kpc)	Linear Diameter M=-0.5 (pc)	R sin β (M=-0.5) (pc)	Notes
103	0 19.8	+60 47	88	+1 1	f	13	0	35	13	12.4	2.40	3.80	14.4	-74	
129	0 24.3	+59 40	88	-2 3	e	11	..	50	13	11.1	1.26	2.00	3.3	-79	
133	0 25.5	+62 48	88	+3 8	e	3	..	50	13	11.7	1.74	2.75	5.6	-40	
146	0 27.5	+62 44	88	+3 8	e	6	Ind	50	10	14.6	0.91	1.45	2.5	-19	
188	0 35.1	+84 47	90	+22 6	c	15	Ind	30	10		6.61	10.5:	46.:	-4000:	1
436	1 9.4	+58 17	94	-3 9	d	4	84	40	12	8.6	0.72	1.15	1.3	-70	
457	1 12.8	+57 48	94	-3 5	e	10	..	100	12	8.6	0.42	0.66	1.9	-45	
559	1 22.8	+62 47	95	+3 3	d	7	..	60	12	11.7	0.74	1.15	5.0	-57	
581	1 26.6	+60 11	96	-1 3	d	5	..	20	12	9.8	0.87	1.38	1.6	-26	
637	1 34.9	+63 32	96	+1 6	d	3	..	20	12	10.2	0.87	1.38	1.2	-63	
654	1 37.2	+61 23	98	-0 1	d	5	..	50	12	9.8	0.72	1.15	1.7	-3	
659	1 37.4	+60 12	98	-1 0	d	5	..	30	12	10.8	1.20	1.91	2.8	-34	
663	1 39.2	+60 44	98	+0 4	e	11	81	80	12	9.6			2.5	-6	
752	1 51.8	+37 11	105	-22 7	f	45	Ind	70		9.5	0.66	1.05	13.8	+405	2
869	2 12.0	+56 41	102	-3 1	f	36	63	..	..	9.5	a	51	26.3	<+136	3
884	2 15.4	+56 39	103	-3 1	e	36	84	..	..	8.6	2		26.3	<+136	3
Mel 15	2 25.2	+56 39	103	0	d	20	..	20	12	11.3	<0.42	<0.66	<3.8	-18	
957	2 26.4	+57 5	104	-2 1	e	10	..	40	12	9.9	1.45	2.29	6.6	+80	
1027	2 35.0	+61 7	103	+2 1	d	7	..	11	12	8.6	<0.72	<1.15	2.3	-43	
1030	2 35.2	+62 51	111	+14 8	e	18	..	80	12	8.6	<0.42	<0.66	<3.5	+168	4
H 1	3 3.1	+62 52	105	+8 0	c	15	Ind	30	12	13.1	<0.42	<0.66	7.9	+166	
1245	3 7.0	+46 46	114	-0 3	e	30	..	40	13.2	9.1	3.	5.0	44.6	+699	
1342	3 25.2	+36 59	123	+5 3	b	15	Ind	40			3.2	5.0	10	+205	
Pleiades	3 41.0	+23 48	134	-22 3	c	120	..	15	12	4.2	0.52	0.83	3.6	+57	5
1502	3 58.6	+62 3	110	+8 4	c	7	..	15		9.8	0.72	1.15	2.3	+169	
1513	4 2.5	+49 15	120	+0 6	c	12	71	40	12	12.8	0.88	1.57	10.3	+46	
1528	4 7.6	+50 59	120	+2 2	c	25	..	80		9.2	0.55	0.87	6.3	+19	
Hyades	4 14.0	+15 23	148	-22 6	c			30		0.7	0.	0.04	10.:	+15	6
1647	4 40.2	+18 53	148	-15 7	c	40	33	40	12	10.9	0.69	1.10	12.8	+288	
1664	4 43.9	+43 31	129	+0 2	e	15	Ind	40		10.9	1.20	1.91	8.3	-24	
1746	4 57.2	+23 40	147	-9 3	e	45	71	60		9.5	0.63	1.00	13.1	+161	
1807	5 9.0	+16 24	153	-12 1	e	10	..	15	12.6	9.7	0.69	1.00	3.1	-230	
1817	5 6.3	+16 34	154	+11 7	d	15	69	10	13	11.1	1.32	2.09	9.1	-486	
1857	5 13.2	+39 14	136	+2 5	d	9		45	12	10.6	1.05	1.66	4.3	+73	

CATALOGUE OF GALACTIC CLUSTERS.—(continued)

N.G.C	R.A. 1900	Dec. 1900	Galactic Long	Galactic Lat.	Class	Approx Diameter	Orientation	Number of Stars	Mag. Lim of Plate	Mag. of 5th Star	Distance $M=+0.5$ (kpc)	Distance $M=-0.5$ (kpc)	Linear Diameter ($M=-0.5$) (pc)	$R \sin \beta$ ($M=-0.5$) (pc)	Notes
H 2	7 51	−25 39	211	+2	d	7	0	20		10.3	0.91	1.45	3.0	+57	
2489	7 52	−29 48	214	+2	g	7	Ind	30	13.5	11.1	1.32	2.09	4.3	+6	
2506	7 55	−10 21	198	+11	g	10	Ind	50	13.8	11.3	1.45	2.29	6.6	+448	
2509	7 56	−18 48	205	+7	g	4	75	40	13.8	10.2	0.87	1.38	1.6	+168	
2516	7 56	−60 36	241	−12	f	60	68	80		10.8	0.83	1.32	23.0	−352	
2530	8 6.0	−12 32	200	+12	f	21		150	12.8	9.4	1.15	1.82	11.2	+390	
2547	8 7	−48 58	233	−9	d	15		80	13.5	9.6	0.60	0.95	4.2	+131	
2548	8 7	−5 30	195	+16	f	30	16	50	13.5	10.0	0.60	0.95	8.3	+271	
2567	8 14	−30 20	217	+3	d	10		30	13.5	9.4	0.79	1.26	3.7	+86	
2571	8 14	−29 26	217	+3	f	8		30	13.5	10.9	0.72	0.95	2.0	+75	
2580	8 17	−30 0	217	+5	c	9		30	13.5	10.9	1.20	1.91	5.0	+154	
2587	8 19	−29 36	217	+5	c	6	69	40	13.5	10.9	1.20	1.91	3.3	+181	
2627	8 33	−29 10	219	+5	c	8			13.5	11.5	1.58	2.51	5.8	+331	
2632	8 34	+20 20	173	+34	d	3		20		12.7	0	18		+10	
I 2391	8 34	−52 34	223	−8	c	30		20	13.5	3.5	0.04	0.06	.38	+363	19
2659	8 37	−44 36	237	−2	d	10		50		12.5	0.51	0.98	0.5	−01	
2660	8 39	−46 51	232	−1	f	5	6	25	13. 4	13.	3.2	5.0	11.6	+69	
I 2395	8 39.3	−48 18	233	−2	d	9	36	30	13. 5	12.2	2.19	3.47	2.2:	+211:	
2670	8 40.0	−47 49	234	−3	f	10		16		10.1	0.83	1.32	0.9	+418	
2671	8 42	−48 25	234	−3	e	15		20		11.2	1.38	2.19	3.8	+68	20
H 3	8 43	−41 31	230	+1	e	3		20		10.4	2	3	0.5	+116	
2682	8 45	+12 11	184	+33	f	7		35	13.4	10.8	0.95	1.51	4.4	+129	
2818	8 12	+36 12	230	+4	f	15	57	67	12.8	13.	1.15	1.82	4.1	+152	
I 2488	9 24	−56 32	245	−8	d	9	33	20	12.5	13	3.2	5.0	7.9	+800:	21
2910	9 26	−52 28	243	−2	f	6		50		13.	3.1	2.19	13:	+156	
2925	9 30	−53 0	243	−1	f	11		30	12.5	12.1	1.38	2.19	8.8	+31	
3105	9 57	−54 38	247		f	5		15	12.5	12.7	1.38	1.74	3.6	+26	
3114	10 1	−59 38	251	+3	e	30	Ind	100		7.	>2.51	>3.98	>17.	+37	
3228	10 17	−51 13	248	+5	f	30		12		10.8	0	0	13:	+19:	
I 2581	10 23	−57 8	252	+1	f	8	31	35		10.1	0.83	2.88	2.6	+105	
3293	10 29	−57 41	252	+1	d	5		50		8.1	0.82	3.2	4.2	+20	
H 4	10 35	−53 38	253	+6	e	8	49	25	12.5	10.2	<0.29	<0.46	<1.0	+0	
Mel 101	10 38	−64 34	257	+4	f	15		40		11.4	0.87	1.38	3.2	+96	
I 2602	10 39	−63 52	257	+5	f	70	49	32		6.1	0.51	1.40	4.1.	+238	22
3532	11 2	−58 8	257	+1	f	60	22	130	.	8.1	0.33	0.52	9.1	+13	

CATALOGUE OF GALACTIC CLUSTERS

CATALOGUE OF GALACTIC CLUSTERS.—(continued)

N G C	R A 1900 (h m)	Dec 1900	Long	Lat	Class	Approx Diameter	Orientation	Number of Stars	Mag of Lum of Plate	Mag of 5th Star	Distance M=+0.5 (kpc)	Distance M=-0.5 (kpc)	Linear Diameter (M=-0.5) (pc)	R sin β (M=-0.5) (pc)	Notes
6208	16 41	−53 38	301 5	+6 8	e	22		50	13 5	12 5	2 51	3 98	2 5	−475	
6222	16 43	−44 33	308 5	+6 3	e	3		20		12 8	2 88	4 57	5	−105:	
6231	16 47	−41 38	311	+1 3	gg	15	44	120		7	<0 24	<0 38	4 0	+4	25
6242	16 48	−39 20	311 5	+0 4	c	10		40	12 2	6 3	<0 36	<0 57	<1 7	+8	26
H12	16 49	−40 33	312 5	+0 2	f	40	45	200	13 2	17 4	<0 38		<4 4	+637	
6253	16 51	−52 33	303 5	+2 7	f	6	49	70		13 3	>3 16	>5 01	>8 7	+168	
6259	16 53	−44 31	309 5	+2 1	e	15		100	12 2	10 4	3 16	2 29	10 0	+2	
6268	16 55	−33 35	314	+5 4	d	10		30		13	1 45	1 51	4 4	−474:	
H13	16 58	−48 45	307	+2 7	f	15		70	13 8	7	0 95	1 50:	22 8	+4	
6281	17 00	−37 45	316	+5 8	g	5	39	25	12 2	11 5	0 21	0 31	0 8	+97	
6318	17 10	−39 40	316	+2 2	c	5		60	12 2	9 6	0 58	0 51	3 0	+79	
6322	17 10	−42 46	313	+4 0	e	12		20	13	10 3	0 66	1 05	5 0	+227	
I4651	17 16	−49 57	307 5	+3 0	f	14		200		11 6	0 66	1 05	2 0	+82	
H14	17 17	−38 57	317	+13 7	g	1		50			1 66	2 63	7 6		
I4757	17 17	−26 15	345 5	+3 1	e	10		2n	18:	8 7					
H15	17 22	−7 26	325	+1 2	c	15	23	15		8 4	0 44	0 69	2 6	+17	27
6383	17 24	−32 46	328	−0 2	c	6	18	20		8 0	0 38	0 60	8	+32	
6400	17 31	−32 30	323	−1 4	e	6	Ind	12		8 1	0 66	1 05	1 2	+24	
6404	17 33	−36 53	320	−2 5	e	3		25	12 2	10 2	0 44	0 69	9	+53	
6405	17 35	−32 09	323	−2 1	g	25	3	50	13	8 6	>0 06	>0 10	>0 4	+46	
H17	17 37	−33 10	324	−6 3	d	20		20		8 3	0 36	0 57	3 1	+20	
6416	17 41	−40 08	317	+15 0	c	60		25	8	7:	0 66	1 05	3 3	+114	
I4665	17 41	+4 5	324 5	+3 0	c	12	23	13		10 3	0 30	0 57	5 2	+29	28
6451	17 44	−30 45	358	−5 6	e	60	18	50		11 8	0 2.	0 3:	10 6	+81.	
6469	17 46	−22 19	327	−5 6	e	12		40	12 2	7 4	0 91	1 45	10 6	+76	
6475	17 47	−34 47	334	−4 7	e	60	44	50	12 2	11 8	1 82	2 88	6 0	+25	29
H18	17 49	−35 16	323 5	+3 7	e	15		86	12 5	7 4	0 24	0 38	3 0	+39	
6494	17 51	−10 0	337 5	−2 8	e	25		120	11:8	10 2	0 87	1 38	10 0	+34	
6520	17 57	−27 54	331	−4 8	g	10	68	25		8 7	0 44	0 69	1 0	+44	30
6531	17 58	−22 30	335	−3 7	e	15		50		9 3	0 57	0 91	2 6	+30	
6530	17 58	−24 20	330	−4 0	f	10		25		9 3	0 57	0 91		+44	
6540	18 00	−27 49	331 5	−1 8	g	1			13	13	3 2:		2 2:		25
6544	18 00	−25 1	333	−2 8	g	0 75		30	13	13			8	+348:	25
6558	18 03	−31 47	327 5		b	5		20	13 2	13 0	3 47	5 50	8	+9	
6583	18 11	−22 10	336 5		e	5	65	50	13 2	13 0	3 16	5 01	5 8	+244	31
6603	18 12	−18 27	340 5		e	4									

6611	18 13.2	−13 49	345	− 0 6	c	25		55	12 5	10 6	1 05	1 66	12 1	− 18	32
6613	18 14 1	−17 10	341 5	− 2 5	d	12		12		10 9	1 20	1 91	6 7	− 83	33
6618	18 15 0	−16 13	342 5	− 2 2	c			35	12 5	10 6	1 05	1 66	..	− 64	34
6633	18 22 7	+ 6 30	3 5	+ 6 8	d	20	.	65	9 6	8 0	0 32.	0 50	2 9·	+ 60·	
I 4725	18 25 8	−19 19	341	− 6 0	d	40	Ind	50	13 2	9.3	0 57	0 91	10 6	− 94	35
6642	18 25 8	−23 32	337	− 7 9	d	1	31	..				.		....	
6645	18 26 8	−16 58	343.5	− 5 1	g	10	14	75	13 2	11 3	1 45	2 29	6 6	− 207	
6649	18 27 9	−10 28	349	− 2 3	f	8	Ind	35	13 2	12 1	2 09	3 31	7 7	− 131	
6664	18 31 1	− 8 18	352	− 2 0	d	18		25	13	11 3	1 45	2 29	12 0	− 79	
I 4756	18 34 0	+ 5 22	4	+ 3 8	d	70		80	10 2	8·	0 3	0 5	10 2	+ 33	
6694	18 39.8	− 9 30	351 5	− 4 4	f	9		20	13	11 9	1 91	3 02	7 9	− 231	36
6705	18 45 7	− 6 23	355	− 4 2	g	10		200	13	12	1 25		3 6	− 92	37
6709	18 46 8	+10 14	10	+ 3 3	d	12		40	13	9 0	0 50	0 79	2 8	+ 46	
6755	19 2 8	+ 4 4	6 5	− 3 1	d	10		50	13	11 0	1 26·	2 00	5 8	− 108	
6756	19 3 8	+ 4 31	6 5	− 3 1	f	3		20		|13	>3 16	>5 01	>4 4	− 272	
6802	19 26 2	−20 4	23 5	− 0 3	e			60		13 0	3 16	5 01.		− 25	
6811	19 35 2	+46 20	47	+11 2	d	15		50		11 8	1 82	2 88	12 6	+ 56	
6819	19 37 9	+39 57	41 5	+ 7 6	d	6		150		13 0	3 16	5 01	8 8·	+ 663.	
6823	19 38 9	+23 4	27	− 1 3	d	5		30	13	9 0	0 50	0 79	1 1.	− 18·	
6830	19 45 8	+22 50	27 5	− 3 0	d	8		20		9 0	0 50	0 79	1 8	− 41	
6834	19 48 2	+29 9	33 5	+ 0 2	d	4		15		13	3 16	5 01	5 8	+ 13	
H 20	19 48 7	+18 5	24	− 5 9	d	10		20		9 0	0 50·	0 79	2 3	− 81	
6838	19 49 3	+18 31	24 5	− 5 8	g	4	Ind	100		11 0	1 26	2 00	2 3	− 202	38
6866	20 0 5	+43 43	47 5	+ 6 0	d	6		50	13	9 4	0 60	0 95	1 7	+ 100	
I 1311	20 7 4	+40 53	35 5	+ 3 4	e	5	.	30		13 0	3 16	5 01	7 3	+ 275	
6885	20 7 8	+26 11	33 5	− 5 1	e	20		35	.	9 8	0 72	1 15	6 7	− 102	
6910	20 19 5	+40 27	46 5	+ 1 3	d	8		40	.	9 0	0 50	0 79	1 8	+ 17	
6913	20 20 3	+38 12	45	− 0 2	d	12		20		9 4	0 60	0 95	3 3	− 4	39
6939	20 29 4	+60 18	63 5	+11 9	g	5	Ind	80		13	3 2	5 0	7 3·	+1034·	
6940	20 30 4	+27 58	37 5	− 8 1	e	20	23	100		11 8	1 82	2 88	16 8	− 406	
7062	21 19 6	+45 57	57 5	− 3 2	e	5		30	13	11 4	1 51	2 40	3 5	− 135	
7086	21 27 1	+51 9	62	− 0 2	e	8		50		11 8	1 82	2 88	6 7	− 10	
7092	21 28 6	+48 0	60 5	− 2 7	c	30		25		6 5	0 16	0 25	2 2	− 12	40
7127	21 40 5	+54 9	65	+ 0 8	d	1		10		11 8	1 82	2 88	0 8	+ 41	
7128	21 40 7	+53 15	64 5	+ 0 1	d	2		20		11 8	1 82	2 88	1 7	+ 4	
7209	22 1 2	+46 0	63	− 7 7	d	20		50		10 1	0 83	1 32	7 7	− 177	
7243	22 11 3	+49 23	66 5	− 5 8	d	20		40		9 0	0 50	0 79	4 7	− 80	
7245	22 11 5	+53 50	69	− 2 1	d	3		40		13	3 16	5 01	4 4	− 183	
7380	22 43 0	+57 34	74 5	− 1 0	d	10		50	13	9 0	0 50	0 79	2 3	− 13	
7419	22 50 4	+60 18	76 5	+ 1 4	e	2		25		13	3 16	5 01	2 9	+ 118	
7510	23 7 2	+60 2	78	+ 0 1	d	2		27		11 7	1 74	2 75	1 6	+ 4	
7654	23 19 8	+61 3	80 5	+ 0 5	e	12	50	120	13	11 0	1 17		4 1	+ 11	41
7762	23 45 0	+67 28	83 5	+ 6 0	d	10		70	13	12 8	2 88	4 57	13 3	+ 480	
H 21	23 49 2	+61 12	83 5	− 0 2	d	5		22		11 7	1 74	2 75	4 0	− 9	
7788	23 51 7	+60 50	83 5	− 0 6	e	2		18		11 0	1 26	2 00	1 2	− 22	
7790	23 52 0	+60 40	83 5	− 0 8	d	6		25		11 7	1.74	2 75	4 8	− 38	
7789	23 52 0	+56 10	83	− 5 2	e	30	Ind	200	13	12 5	2 51	3 98	34 8	− 362	

NOTES TO APPENDIX B

[1] Messier 103

[2] Distance from Wallenquist, Ups. Medd. 42, 1929

[3] h and χ Persei; fifth star assumed of magnitude $-2^m.5$.   Trumpler (P. A. S. P , 38, 352, 1926) gives a distance of 2.3 on the basis of the mean magnitude for Class Ao

[4] Messier 34

[5] Messier 45; the diameter from Russell, Dugan, and Stewart corresponds to an angular radius of nearly two degrees

[6] Distance and diameter from Russell, Dugan, and Stewart

[7] Messier 38

[8] Messier 36; distance from Wallenquist,Ups. Medd. 32, 1927

[9] Messier 37; distance from von Zeipel and Lindgren, Swedish Acad., 61, No. 15, 1921

[10] Messier 35

[11] Scarcely resolved

[12] Eccentric in an elliptical diffuse nebula

[13] S Monocerotis in cluster

[14] Messier 41

[15] Messier 50

[16] τ Canis Majoris in cluster

[17] Messier 46; an important photographic catalogue by Chevalier (see App. C)

[18] Messier 93

[19] Praesepe, Messier 44; distance from Russell, Dugan, and Stewart

[20] Messier 67

[21] Diffuse nebula in a coarse cluster

[22] θ Carinae involved

[23] Coma Berenices; distance from Russell, Dugan, and Stewart

[24] κ Crucis

[25] Not resolved; a photograph of N. G. C. 6540 appears in Pop. Ast. Tidsk., 8, 62, 1927

[26] Mount Wilson plate

[27] Messier 6

[28] Messier 7

[29] Messier 23

[30] Messier 21

[31] Messier 24

[32] Messier 16

[33] Messier 18

[34] Messier 17; nebulosity

[35] Messier 25

[36] Messier 26; Trumpler finds a distance of 2.75 (L. O B., 14, 122, 1929)

[37] Messier 11; distance from Trumpler

[38] Messier 71

[39] Messier 29

[40] Messier 39

[41] Messier 52; distance from Wallenquist, Ups. Medd. 42, 1929

APPENDIX C

BIBLIOGRAPHY

THE bibliography of star clusters contains only a few entries prior to 1875 but is essentially complete from then to 1930. Bibliographies of earlier papers have been compiled by Holden (Reference 261 below) and Knobel (Reference 316). Some papers on Cepheid variables, mainly my own, are included in the bibliography because of their bearing on the period-luminosity and period-spectrum relations which have been of importance in cluster work. A few articles on the Magellanic Clouds and the structure of the Galaxy are also listed.

The topical bibliographies in Appendix D should facilitate study in a few special fields.

1. ADAMS, W. S., *The Radial Velocities of the Brighter Stars in the Pleiades*, Ap. J., 19, 338, 1904.
2. ———— and A VAN MAANEN, *A Group of Stars of Common Motion in the h and χ Persei Clusters*, A. J., 27, 187, 1913.
3. ————, *The Spectra of Some Individual Stars in the Hercules Cluster*, P. A. S. P , 25, 260, 1913.
4. ———— and H. SHAPLEY, *The Spectrum of δ Cephei*, Mt. W. Comm 22, 1916.
5. ———— and ————, *Note on the Cepheid Variable SU Cassiopeiae*, Mt. W. Contr. 145, 1918.
6. ALDEN, H. L., *Trigonometric Parallax of the Pleiades*, Amer. Astr. Soc , 4, 349, 1922.
7. ANDREINI, A., *Distanze e Dimensioni cosmiche*, Livorno, 1921.
8. AUWERS, A., *Aus einem Schreiben des Herrn Geheimrath Auwers . . .* , A. N., 114, 47, 1886 (Nova in Messier 80).
9. BAADE, W., *Sieben Veränderliche in der Umgebung des Kugelhaufens M 53*, Hamburg Mitteilungen, 5, No. 16, 1922.
10. ————, *Bemerkung zu der Arbeit von H. Kienle und P. ten Bruggencate uber die absolute Helligkeit der Plejadensterne*, Zeit. f. Phys., 31, 604, 1924.
11. ————, *5 isolierte Haufenveränderliche in der Umgebung des Kugelhaufens N. G. C. 5466*, Hamb. Mitt., 6, No. 27, 1926.
12. ————, *Der kugelförmige Sternhaufen N. G. C. 5466*, Hamb. Mitt., 6, No. 27, 1926.
13. ————, *17 neue Veränderliche im Kugelhaufen M 53 (N. G. C. 5024)*, Hamb. Mitt., 6, No. 27, 1926.

14. ———, *Der Sternhaufen N. G. C.* 5053, Hamb. Mitt., **6**, No. 29, 1928; A. N., **232**, 193, 1928. (See Chapter II, Section 8 above for a discussion of this abnormal cluster.)

15. BACKHOUSE, I. W., *Note on Mr. A. S. Eddington's Moving Cluster of Stars in Perseus*, M. N. R. A. S., **71**, 523, 1911.

16. BAILEY, S. I , *ω Centauri*, Astr and Ap., **12**, 689, 1893.

17. ———, *Variable Stars in Clusters (Abstract)*, Amer. Astr. Soc., **1**, 49, 1898.

18. ———, *The Periods of the Variable Stars in the Cluster Messier* 5, Ap. J., **10**, 255, 1899.

19. ———, *Note on the Relation between the Visual and Photographic Light Curves of Variable Stars of Short Period*, Ap. J., **10**, 261, 1899.

20. ———, *The Rate of Increase in Brightness of Three Variable Stars in the Cluster Messier* 3, Amer. Astr. Soc., **1**, 100, 1900; Science, **12**, 122, 1900

21. ———, *ω* Centauri, H. A., **38**, 1, 1902.

22. ———, *Variable Stars in the Clusters Messier* 3 *and Messier* 5, H. C. 100, 1905.

23. ———, *The Number and Distribution of Stellar Clusters and Nebulae*, Amer. Astr. Soc , **1**, 268, 1906.

24. ———, *Note on the Magnitude of the Stars in Messier* 3 *(Abstract)*, J. R.-A. S. Can., **5**, 337, 1911.

25. ———, *Variable Stars in the Cluster Messier* 3, H. A., **78**, 1, 1913.

26. ———, *Variable Stars in the Cluster Messier* 5, H. A., **78**, 103, 1917.

27. ———, *Variable Stars in the Cluster Messier* 15, H. A., **78**, 199, 1919.

28. ———, *Globular Clusters*, VJS. d. A G., **48**, 418, 1913.

29. ———, *On the Number of the Globular Clusters (Abstract)*, Pop. Astr., **22**, 558, 1914.

30. ———, *Globular Clusters*, H. A., **76**, No. 4, 1915.

31. ———, *Cluster Variables with Double Maxima*, H. C. 193, 1916.

32. ———, *Note on the Form of the Light Curve of Variable Stars of Cluster Type*, *(Abstract)*, Pop. Astr., **25**, 307, 1917.

33. ———, *Note on the Variable Stars in the Globular Cluster M* 15 *(Abstract)*, Pop. Astr., **25**, 520, 1917.

34. ———, *Note on the Magnitudes of the Variables in M* 15 *(Abstract)*, Pop. Astr , **26**, 683, 1918.

35. ———, *Globular Clusters*, H. C. 211, 1918.

36. ———, *Variable Stars in M* 22 *(Abstract)*, Pop. Astr., **28**, 518, 1920.

37. ———, *Variable Stars in the Cluster N. G. C.* 6723, H C. 266, 1924.

38. ———, *Photographic Work at Arequipa with the Bruce* 24-*inch Refractor, N. G. C.* 3201, H. C. 234, 1922 (Variable Stars in N. G. C. 3201).

39. ———, *Eight New Variable Stars near N. G. C.* 6809, H. B. 813, 1925.

40. ———, *Clusters and Nebulae*, H. Repr. 29, 1926.

41. BALANOWSKY, J., *(Photometric Study of* 42 *Stars in the Cluster h Persei)*, Pulk. Bul., **7**, 199, 1920 (in Russian).

42. ———, *Étude des amas stellaires h et χ Persée*, Pulk. Bul., **9**, 277, 1924.

43. ———, *Die Eigenbewegung des kugelförmigen Sternhaufens Messier* 92 *(N. G. C.* 6341), Pulk. Bul., **11**, 167, 1928; C. R. Acad. U. R. S. S., No. 21, 364, 1927.

44. BALL, R., and E. A. RAMBAUT, *On the Relative Position of 223 Stars in the Cluster χ Persei*, Trans. Roy. Irish Acad., 30, Part 4, 1892.

45. BANNISTER, R. D., *Positions of the Stars in a Cluster Whose Center Is at R. A. 18ʰ34ᵐ and Declination +5°17' Determined from Photographic Plates*, A. J., 31, 165, 1918. (This cluster is I. C. 4756.)

46. BARABASCHEFF, N., *Über die Helligkeitsverteilung im Sternhaufen M 13*, A. N., 220, 299, 1923.

47. BARNARD, E. E., *The Cluster G. C. 1420 and the Nebula N. G. C. 2237*, A. N., 122, 53, 1889.

48. ———, *Photographic Nebulosities and Star Clusters connected with the Milky Way*, Astr. and Ap., 13, 177, 1894.

49. ———, *Triangulation of Star Clusters*, Amer. Astr. Soc., 1, 77, 1899.

50. ———, *Note on Some of the Variable Stars of the Cluster Messier 5*, A. N., 147, 243, 1898.

51. ———, *Triangulation of Star Clusters*, Science, 10, 789, 1899.

52. ———, *Note on the Exterior Nebulosities of the Pleiades*, M. N. R. A. S., 59, 155, 1899, 60, 258, 1900.

53. ———, *Some Abnormal Stars in the Cluster M 13 Herculis*, Ap. J., 12, 176, 1900.

54. ———, *Micrometrical Measures of Individual Stars in the Great Globular Clusters*, Science, 17, 330, 1903

55. ———, *Discovery and Period of a Small Variable Star in the Cluster M 13 Herculis*, Ap. J., 12, 182, 1900 (Previously discovered by Bailey; cf. Barnard, Ap. J., 29, 72, 1908).

56 ———, *Micrometrical Measures of Individual Stars in the Great Globular Clusters*, Amer. Astr. Soc., 1, 193, 1902.

57. ———, *On Some of the Variable Stars in the Cluster M 5 Librae*, Amer. Astr. Soc., 1, 193, 1902.

58. ———, *Visual Observations of a Variable Star in the Cluster M 3 (N. G. C. 5272)*, A. N., 172, 345, 1906.

59. ———, *On the Motion of the Stars in the Cluster Messier 92*, A. N., 176, 17, 1907.

60. ———, *Second Paper on the Motion of the Stars in Messier 92*, A. N., 176, 21, 1907.

61 ———, *On the Colors of Some of the Stars in the Globular Cluster M 13 Herculis*, Ap. J., 29, 72, 1908.

62. ———, *On the Constancy of the Period of the Variable Star, M 5 (Librae) No. 33*, Amer Astr. Soc., 1, 298, 1908.

63. ———, *On the Proper Motion of Some of the Small Stars in the Dense Cluster M 92 Herculis*, Amer. Astr. Soc., 1, 323, 1909.

64. ———, *On the Motion of Some of the Stars of Messier 92 (Hercules)*, A. N., 182, 305, 1909; Pop. Astr., 18, 3, 1910.

65. ———, *On the Period and Light Curve of the Variable Star No. 33, M 5 (Libra) and on the Possible Use of Such a Star as a Time Constant*, A. N., 184, 273, 1909.

66. ———, *The Variable Star No. 33 in the Cluster M 5*, A. N., 196, 11, 1913.

67. ———, *On the Change in the Period of the Variable Star Bailey No. 33 in the Cluster M 5 (Abstract)*, Pop. Astr, 27, 522, 1919.

68. ———, *Remeasurement of Hall's Stars in the Pleiades (Abstract)*, Pop. Astr., **27**, 523, 1919.

69. ———, *Variable Stars in the Cluster M 11 (N. G. C. 6705)*, Pop. Astr., **27**, 485, 1919.

70. ———, *On the Comparative Distances of Certain Globular Clusters and the Star Clouds of the Milky Way*, A. J., **33**, 86, 1920.

71. BATTERMANN, H., *Triangulation zwischen den acht hellsten Sternen der Plejadengruppe*, A. N., **122**, 353, 1889.

72. BAXENDELL, J., *Notes on Pogson's Observations of U Geminorum, T Scorpii, and R Librae*, A. J., **22**, 127, 1902 (Nova in Messier 80).

73. BECKER, F., *Die Kugelformigen Sternhaufen*, Himmelswelt, **32**, 105, 1922.

74. ———, *Sternhaufen und Nebelflecken*, Hevelius, p. 354, 1922.

74a ———, *Die Verteilung der Spektren in zwei Offenen Sternhaufen*, A. N., **236**, 327, 1929.

75. BELLAMY, F. A., *A New Cluster in Cygnus, with Right Ascensions and Declinations of 103 Stars Included in It*, M. N. R. A. S., **64**, 662, 1904.

76. BELOPOLSKY, A., *Über die Veranderungen in dem Sternhaufen N. G. C. 5272*, A. N., **140**, 23, 1895.

77. BELOT, E., *L'Origine Possible des Amas d'Etoiles*, C. R., **164**, 513, 1917.

78. BERGSTRAND, Ö., *Sur le Groupe des Etoiles à Hélium dans la Constellation d'Orion*, N. Acta S. S. U., Reg. Soc. Sci. Ups., Ser. 4, Vol. 5, No. 2, 1919.

79. BHASKARAN, T. P., *Proper Motions of Stars in the Cluster M 41 (N. G. C. 2287)*, M. N. R. A. S., **79**, 59, 1918.

80. BICKERTON, A. W, *The New Astronomy, III. Star Clusters and Nebulae*, Knowledge, **34**, 413, 1911.

81. BIGOURDAN, G., *Les Nébuleuses de la Région des Pléiades*, Bul. Astr., **28**, 417, 1911 (History of discoveries, and description of the nebulosities, rather extensive bibliography).

82. ———, *Observations de Nébuleuses et d'Amas Stellaires*, Ann. Obs. Paris, Observations, 1884, 1888, 1890–1907.

83. BOHLIN, K., *Der Zweite Sternhaufen im Hercules, Messier 92*, Stockholm Publ., **8**, No. 3, 1906.

84. ———, *Ausmessung des Zweiten Sternhaufens im Hercules (Messier 92)*, A. N., **174**, 203, 1907.

85. ———, *On the Galactic System with Regard to Its Structure, Origin, and Relations in Space*, Swedish Acad. Proc., **43**, No. 10, 1909.

86. BOS, W. H. VAN DEN, *Photovisual Magnitudes of 55 Stars in Praesepe, from Plates taken at Potsdam*, B. A. N., **1**, 79, 1922.

87. BOSS, B, *Systematic Proper-Motions of Stars of Type B*, A. J, **26**, 163, 1910. (See also A. J., **27**, 33, 1911; **27**, 67, 1912; **28**, 12, 15, 1913; **28**, 174, 1914.)

88. ———, *A Convergent Point for Four Clusters of Small Proper Motion Stars*, A. J., **30**, 22, 1916.

89. BOSS, L., *Convergent of a Moving Cluster in Taurus*, A. J., **26**, 31, 1908.

90. BOTTLINGER, K. F., *Die Eigenbewegung der Bärengruppe*, A. N., **198**, 153, 1914.

91. BOURGET, H., *Photographie des Nébuleuses et des Amas stellaires*, Bul. Soc. Astr. France, **14**, 57, 1900.

92. BREDICHIN, T., *Mesures micrométriques du groupe de Persée*, Ann. Moscow Obs., **4**, No. 2, 1878.

93. BRONSKY, M., and A. STEBNITZKY, *Les Positions des Etoiles de h et χ Persée et de leurs Environs*, Mem. St. Petersburg Acad. Sci., **2**, Ser. 8, No. 7, 1894.

94. BROWN, F.L , *Measures of the Cluster N.G C.* 6633, A. J., **31**, 57, 1918.

95. BRUGGENCATE, P. TEN, *Die Bedeutung der Farbenhelligkeits-diagrammen für das Studium der Sternhaufen*, Probleme der Astronomie, Festschrift für Hugo von Seeliger, p. 50, 1924.

96. ———, *Uber die Reste einer Spiralstruktur in Sternhaufen*, Zeit. f. Phys., **24**, 48, 1923.

97. ———, *Uber eine Absorption des Lichtes bei offenen Sternhaufen*, Zeit. f. Phys. **29**, 243, 1924.

98. ———, *The Absorption of Light in Open Star Clusters*, B. A. N., **4**, 51, 1927.

99. ———, Sternhaufen, Berlin, 1927.

100. ———, *On the Determination of the Spacial Distribution of the Stars in Globular Clusters from the Intensity Distribution in Their Projections*, B. A. N , **4**, 195, 1928.

101. ———, *Bemerkungen uber ellipsoidformige Sternhaufen*, A. N., **232**, 417, 1928.

102. ———, *Die Dichteverteilung in rotationssymmetrischen Sternhaufen*, A. N., **232**, 423, 1928.

102a ———, *The Radial Velocities of Globular Clusters*, P. N. A. S., **16**, 111, 1930.

103. CERASKI, W., *Über die Anzahl der Sterne in den Plejaden*, A. N., **108**, 245, 1884.

104. ———, *Etude Photométrique sur l'Amas Stellaire Coma Berenices*, Moscow Annals, Ser. 2, **6**, 33, 1917.

105. ———, *Une nouvelle Variable 16.1904 Persei au Cluster χ Persei*, A. N., **165**, 126, 1904.

106. CHAPMAN, S., *Some Problems of Astronomy, II. Globular Clusters*, Obs., **36**, 112, 1913.

107. CHARLIER, C. V. L., *Preliminary Statistics of Nebulae and Clusters*, Lund Medd. 56, 1913. (Arrays and diagrams based on the N. G. C. and the I. C.; correlations of distribution, brightness, and size.)

108. ———, *Stellar Clusters and Related Celestial Phenomena*, Lund Medd., Ser. 2, 19, 1918.

109. CHASE, F. L., *Triangulation of the Principal Stars of the Cluster in Coma Berenices*, Trans. Yale Obs., **1**, Part V, 1896.

110. CHASE, H. S., *A Comparison of the Positions of the Stars in Praesepe Derived by Dr. B. A. Gould from Photographs with the Positions Observed by Professor Hall*, A. J, **8**, 167, 1889

111. CHEVALIER, S., *Etude Photographique de l'Amas d'Etoiles Messier 67 (N. G. C.* 2682), Ann. Zô-Sè Obs , **8b**, 1914.

112. ———, *Etude Photographique de l'Amas d'Etoiles Messier 46 (N. G. C.* 2437), Ann. Zô-Sè Obs., **9d**, 1, 1916.

113. ———, *Amas d'Etoiles Messier 22 (N. G. C.* 6656), Ann. Zô-Sè Obs., **100**, 1, 1918.

114. CHÈVREMONT, A., *Découverte d'une Etoile variable dans l'Amas Messier 2 du Verseau*, Bul. Soc. Astr. France, **12**, 16, 90, 1898.

115. CHRÉTIEN, H., *Sur l'Analyse statistique des Amas d'Etoiles*, C. R., **157**, 1047, 1913.

116. CLERKE, A. M., *The Yale College Measurement of the Pleiades*, Nature, **36**, 372, 1887.

117. COEBERGH, C. L. A. M., *Parallaxbepalingen van Sterrenhoopen*, Hem. en Damp , 1918.

118. COLLINDER, P., *Sur la Distribution et les Couleurs des Etoiles dans l'Amas M 34*, Ark. Mat. Astr. Fys., **19B**, No. 11, 1926.

119. ——, *De öppna Stjärnhoparna i Vintergatan och i Angränsande Stjärnsystem*, Pop. Astr. Tid., **8**, 36, 1927.

120. COMMON, A. A., *Faint Stars near Alcyone*, M. N. R. A. S., **44**, 412, 1884.

121. ——, *The Photographic Nebulae in the Pleiades*, M. N. R. A. S., **46**, 341, 1886.

122. ——, *Note on some Variable Stars near the Cluster 5 M*, M. N. R. A. S., **50**, 517, 1890.

123. COMSTOCK, G., *A New Member of the Taurus Cluster*, A. J., **34**, 33, 60, 1922.

124. CROMMELIN, A. C. D., *The Distances of Globular Clusters*, Obs., **45**, 138, 1922.

125. CURTIS, H. D., *Descriptions of Nebulae and Clusters Photographed with the Crossley Reflector*, L. O. B., **7**, 81, 1912; **8**, 43, 1913.

126. ——, *Search for Faint Members of the Taurus Cluster*, P. A. S. P., **27**, 243, 1915.

127. ——, *Finding List for General Catalogue Numbers*, P. A S P., **29**, 180, 1917.

128. ——, *Descriptions of 762 Nebulae and Clusters Photographed with the Crossley Reflector*, Lick Publ., **13**, 9, 1918.

129. ——, *The Scale of the Universe*, Bul. Nat. Res. Coun., **2**, 194, 1921.

130. DAVIS, H., *A Bright Variable Star in N. G. C. 6779 (M 56)*, P. A. S. P., **29**, 210, 1917.

131. ——, *Five new Variable Stars in Globular Clusters*, P. A. S. P., **29**, 260, 1917.

132. DAWSON, B. H., *Connections of Cluster Stars*, La Plata Results, **4**, Part Ia, 125, 1918; **4**, Part II, 260, 1922.

133. D'ESTERRE, C R., *A Note on some Observations of the Region Around the Star Clusters H. VI. 33, 34, Persei*, M. N. R. A. S., **73**, 75, 1912.

134. DENNING, W. F., *The Cluster No. 361, Dreyer's Index Catalogue (1895)*, Obs., **41**, 140, 1918.

135. DOIG, P., *Note on the Parallaxes of Open and Moving Clusters*, M. N. R. A. S., **82**, 461, 1922.

136. ——, *The Distribution of Certain Celestial Objects in Galactic Longitude*, J. B. A. A., **33**, 238, 1923.

137. ——, *An Estimate of the Distances of 14 Open Clusters*, J. B. A. A., **35**, 201, 1925 (Spectral parallaxes).

138. ——, *Note on the Distance of M 11 Aquilae (N. G. C. 6705)*, J. B. A. A., **35**, 237, 1925.

139. ——, *The Spacing of the Nearby Stars compared with Globular Clusters*, J. B. A. A , **35**, 288, 1925.

140. ——, *Note on the Nebulae and Clusters in Webb's "Celestial Objects for Common Telescopes,"* J. B. A. A., **36**, 89, 1926.

141. ——, *A Catalogue of Estimated Parallaxes of 112 Nebulae, Open Clusters, and Star Groups,* J. B. A. A., **36**, 107, 1926.

142. ——, *Spectral Types in Open Clusters,* P. A. S. P., **38**, 113, 116, 1926.

143. ——, *The Positions in Space of 76 Open Clusters and Star Groups,* J. B. A. A., **36**, 115, 1926.

144. ——, *The Physical Connection between the Stars of a Group or Cluster,* J. B. A. A., **36**, 289, 1926.

145. ——, *Sir William Herschel's Estimates of Globular Cluster Distances,* J. B. A. A., **37**, 99, 1927.

146. ——, *The Luminosities of Cepheids,* Obs., **51**, 197, 1928; See, also, J. B. A. A., **38**, 255, 1928.

147. DONNER, A. AND O. BACKLUND, *Positionen von 140 Sternen des Sternhaufen 20 Vulpeculae,* Bul. de l'Acad. Imp. des Sci. de St. Petersbourg, Ser 2, **5**, No. 2, 77, 1895.

148. DREYER, J. L. E., *A New General Catalogue of Nebulae and Clusters of Stars,* Mem. R. A. S., **49**, Part I, 1888.

149. ——, *Index Catalogue of Nebulae found in the Years 1888–1894,* Mem. R. A. S., **51**, 1895.

150. ——, *Second Index Catalogue of Nebulae and Clusters of Stars,* Mem. R. A. S., **59**, Part II, 1908.

151. ——, *Corrections to the New General Catalogue, Resulting from the Revision of Sir William Herschel's Three Catalogues of Nebulae,* M. N. R. A. S , **73**, 37, 1912.

151a. DUFAY, J, *Les Grandeurs integrales et les Distances relative des Amas globulaires,* Bull. de l'Observatoire de Lyon, **11**, 59, 1929.

152. DUNCAN, J. C., *Bright Nebulae and Star Clusters in Sagittarius and Scutum Photographed with the 60-inch Reflector,* Mt. W. Contr. 177, 1919.

153. DUGAN, R. S., *Helligkeiten und Mittlere Oerter von 359 Sternen der Plejadengruppe,* Heidelberg Ap Publ., **2**, 29, 1905.

154. DZIEWULSKI, W., *Uber die Bewegung einiger Sterngruppen im Raume,* Bul. de l'Akad. des Sci. de Cracovie, A, Math. Kl., 185, 1915.

155. ——, *Über die Bewegung des Systems der Sterne α Lyrae, τ Coronae Borealis, ε Cephei,* Extrait du Bul. de l'Acad. des Sci. de Cracovie, Math. Kl., 50, 1916 (In Polish).

156. ——, *Kritische Bemerkungen uber Sterngruppen, mit Berücksichtigung der Sterne der Gruppe β, γ, δ, ε, ζ Ursae Majoris,* Bul. de l'Acad. des Sci. de Cracovie, Math. Kl., 251, 1916.

157. EDDINGTON, A. S., *Note on a Moving Cluster of Stars of the Orion Type in Perseus,* M. N. R. A. S., **71**, 43, 1910.

158. ——, *The Dynamics of a Globular Stellar System,* M. N. R. A. S., **74**, 5, 1913; **75**, 366, 1915; **76**, 37, 1915.

159. ——, *The Kinetic Energy of a Star Cluster,* M. N. R. A. S., **76**, 525, 1916.

160. ——, *The Distribution of Stars in Globular Clusters,* M. N. R. A. S., **76**, 572, 1916.

161. ——, *The Nature of Globular Clusters,* Obs., **39**, 513, 1916.

162. ——, *Researches on Globular Clusters*, Obs., **40**, 394, 1917 (summary of Shapley's work).

163. ——, *The Distribution of Globular Clusters (Council Note)*, M. N. R. A. S., **79**, 292, 1919.

164. EINSTEIN, A., *Eine einfache Anwendung des Newtonschen Gravitationsgesetzes auf die kugelförmigen Sternhaufen*, Festschrift Kais. Wil. Ges., p. 50, 1921.

165. ELKIN, W. L., *Determination of the relative Positions of the Principal Stars in the Group of the Pleiades*, Yale Obs. Trans., I, 1, 1887.

166. ——, *Comparison of Dr. Gould's Reductions of Mr. Rutherfurd's Pleiades Photographs with the Heliometer-Results*, A. J., **9**, 33, 1889.

167. ——, *The Rutherfurd Photographic Measures of the Pleiades*, P. A. S. P., **4**, 134, 1892.

168. ——, *Revision of the First Yale Triangulation of the Principal Stars in the Group of the Pleiades*, Yale Obs. Trans., I, Part 7, 331, 1896.

169. ELLERY, R. L. J, *Photograph of κ Crucis*, M. N. R. A. S., **43**, 395, 1883.

170. ENGELHARDT, *Notiz zu "Muthmassliche starke Eigenbewegung eines Sterns im Sternhaufen G. C. 4440,"* [1] A. N., **120**, 39, 1889.

171. ENGELMANN, R., *Meridianbeobachtungen von Nebelflecken*, A. N, **104**, 193, 1882. (Some globular clusters included.)

172. ESPIN, T., *The Red Stars in the Great Perseus Cluster*, M. N. R. A. S, **52**, 154, 1892.

173. FAGERHOLM, E., *Photographical Measurement of the Principal Stars in the Cluster of Coma Berenices and Determination of their Proper Motions*, Ark. Mat. Astr. o Phys, **2**, No. 31, 1906

174. ——, *Über den Sternhaufen M 67*, Dissertation, Upsala, 1906.

175. ——, *Undersökningar öfver stjärnhopen G C. 341, I. Fotografisk uppmätning*, Ark. Mat. Astr. o. Phys., **5**, No. 14, 1909.

176. FAIRFIELD, P., *Proper Motion of N. G. C. 6231 = Dunlop 499*, H. B. 843, 1927.

177. FATH, E. A., *The Spectra of some Spiral Nebulae and Globular Star Clusters*, L. O. B. 149, 1909.

178. ——, *The Distribution of Nebulae and Globular Star Clusters*, Pop. Astr, **18**, 544, 1910.

179. ——, *The Spectra of Spiral Nebulae and Globular Star Clusters. Second Paper*, Mt. W. Contr 49, 1911.

180. ——, *The Spectra of Spiral Nebulae and Globular Star Clusters. Third Paper*, Mt. W. Cntr. 67, 1913.

181. FENET, L., *L'Amas Messier 11 de l'Aigle*, Bul. Soc. Astr. France, **9**, 83, 1895.

182. FLAMMARION, C., *Photographie des Pléiades*, Bul. Soc. Astr. France, **7**, 193, 1893.

183. ——, *Nébuleuses et Amas d'Etoiles de Messier; Observations méthodiques faites à l'Observatoire de Juvisy*, Bul. Soc. Astr. France, **31**, 385, 1917; **32**, 25, 56, 98, 123, 160, 196, 239, 267, 308, 340, 402, 446, 1918; **33**, 21, 79, 169, 206, 263, 318, 341, 383, 415, 455, 517, 1919.

184. ——, *Les Principaux Amas d'Etoiles et Grandes Nébuleuses du Ciel visible, en France*, Bul. Soc. Astr. France, **34**, 34, 62, 132, 164, 222, 273, 323, 366, 416, 455, 498, 534, 1920; **35**, 22, 60, 112, 143, 193, 243, 287, 331, 355,

377, 429, 460, 1921; 36, 23, 70, 113, 207, 271, 298, 403, 477, 1922; 37, 113, 262, 458, 1923.

185. FLEMING, W. P., *Note on Mr. Packer's Variables near M 5 Librae*, Sid. Mess , 9, 380, 1890.

186. ———, *Two New Variable Stars near the Cluster 5 M Librae*, A. N., 125, 157, 1890.

187. FRANZ, J., *Uber die Nebelflecken bei den Plejaden*, Schles. Ges. f. vaterl. Cult , 78, Part 2, 36, 1900.

188. FREUNDLICH, E., *Zur Dynamik der Kugelförmigen Sternhaufen*, Phys Zeit., 24, 221, 1923.

189. ——— AND V. HEISKANEN, *Über die Verteilung der Sterne verschiedener Massen in den kugelformigen Sternhaufen*, Zeit. f. Phys., 14, 226, 1923.

190. GAULTIER, E. C., *Catalogue Annuel des Grandeurs Photographiques de 300 étoiles des Plétades*, Bul. Soc. Astr. France, 15, 491, 1901.

191. GIEBELER, H , *Der Sternhaufen M 37 nach photographischen Aufnahmen am Bonner Refraktor von F. Küstner und Ausmessungen von H. Stroele*, Bonn Veröff., No. 12, 1914.

192. GIFFORD, A. C., *The Average Distance Apart of Stars in a Globular Cluster*, J. B. A. A., 36, 31, 1926.

193. GORE, J. E., *Globular Star Clusters*, Knowledge, 17, 232, 255, 1894.

194. ———, *Messier's Nebulae*, Obs., 25, 264, 288, 321, 1902.

195. GOTHARD, E. VON, *Muthmassliche starke Eigenbewegung eines Sterns im Sternhaufen G. C. 4440*, A. N., 116, 257, 1887.

196. GOULD, B. A , Photographic observations of star clusters, 1897.

197. GRAFF, K., *Photometrische Durchmusterung der Plejaden bis zu Sternen 14. Grösse*, Hamburg Abhandlungen, 2, 3, 1920.

198. ——— AND W. KRUSE, *Photometrische Vermessung des Sternhaufens N. G. C. 6633*, A. N., 214, 171, 1921.

199. ———, *Photometrische Helligkeiten und Farben in dem Sternhaufen M 34 Persci*, A N., 219, 297, 1923.

200. ———, *Photometrische Stern- und Farbenfolge in dem zerstreuten Sternhaufen N. G. C. 7209*, A. N., 223, 161, 1924.

201. GUSHEE, V. M., *A Study of Proper Motions in the Cluster N. G. C. 663*, A. J., 32, 117, 1919.

202. GUTHNICK, P., *Kugelhaufen, inbesondere über gemeinsam mit Herrn R. Prager begonnene Untersuchungen an M3, M13, M15, und M92 (Abstract)*, Sitz. d Preuss. Akad. d. Wiss., 27, 508, 1925.

203. ——— AND R. PRAGER, *Uber eine Gesetzmässigkeit im System Ursa Major*, Arb. aus den Gebeiten d. Phys., Math., und Chem., Braunschweig (Pub Vieweg), 652, N. D.

204. GYLLENBERG, W., *Die Ausmessung des Sternhaufens I. C. 4996*, Lund Medd. 105, 1925.

205. HAGEN, J. G., *On the Extension and Appearance of N. G. C. 6822*, Atti Pont. Acc. Nuovo Lincei, 77, 135, 1924.

206. HAHN, R., *Mikrometrische Vermessung des Sternhaufens Σ 762*, Abh. d. Math.-Phys. Kl. d. Kgl Sächs. Ges. d. Wiss., 17, 151, 1891.

207. ——, *Mikrometrische Vermessung des Sternhaufens* Σ 762, A. N., **129**, 395, 1892.

208. HALE, G. E., *Photographs of Star Clusters Made with the Forty-inch Visual Telescope*, Ap. J., **12**, 161, 1900; Yerkes Obs. Bul. 15, 1900.

209. HALL, A., *Relative Positions of 63 Small Stars in the Pleiades*, A. J., **7**, 73, 1887.

210. HARTMANN, J., *Die Bewegung der elf hellsten Plejadensterne*, A. N , **199**, 305, 1914.

211. HAYN, F., *Eigenbewegungen und Parallaxe der Plejaden*, A. N., **198**, 147, 1914.

212. ——, *Katalog von 70 Plejadensternen fur das Aquinoktium von* 1900.0, A. N., **209**, 355, 1919.

213. ——, *Nachtrag zum Katalog der Plejaden in A. N.*, **209**, 355, A. N., **211**, 233, 1920.

214. ——, *Die Plejaden*, Abh. der Math.-Phys. Kl. d. Sachs. Akad. d. Wiss., **38**, No. 6, 1921.

215. ——, *Der Sternhaufen Praesepe*, Leipzig Veröff., **2**, 1927.

216. HECKMANN, O., *Photographische Vermessung der Praesepe*, A. N., **225**, 49, 1925.

217. ——, *Analysis of ten Bruggencate's "Sternhaufen,"* VJS. d. A. G., **62**, 180, 1927.

218. ——, *Photographische Vermessung der Sterngruppe Coma Berenices*, Gött Veröff , **5**, 1929.

219. —— and H. SIEDENTOPF, *Über die Struktur der Kugelformigen Stern-haufen*, Gött. Veröff., **6**, 1929; Zs. f. Phys., **54**, 183, 1929

220. HEILMANN, J., *Berichtigung*, A. N., **236**, 211, 1929. (Concerns Messier 11 and the Scutum star cloud.)

221. HEINEMANN, K., *Photographische Photometrierung und Vermessung des Haufens N. G. C.* 752, A. N., **227**, 193, 1926.

222. HELFFRICH, J., *Untersuchungen im Sternhaufen h Persei*, Heidelberg Veröff., **7**, No. 2, 29, 1913.

223. HELMERT, F. R., *Der Stern-Haufen im Sternbilde des Sobieski'schen Schildes*, Hamburg Publ. No. 1, 1874.

224. HENKEL, F. W., *Clusters and Nebulae*, Knowledge, **34**, 342, 1911.

225. HENRY, M M., (χ *Persei*), Sirius, **18**, 1885, Plate 11.

226. ——, *The Photographic Nebulae in the Pleiades*, M. N. R. A. S., **46**, 98, 281, 1886; Bul. Soc. Astr. France, **2**, 106, 1888.

227. HERTZSPRUNG, E., *On New Members of the System of the Stars β, γ, δ, ε, ζ, Ursae Majoris*, Ap. J., **30**, 135, 1909; (correction p. 320, 1909).

228. ——, *Über die Verwendung photographischer effektiver Wellenlängen zur Bestimmung von Farbenäquivalenten*, Pots. Publ., **22**, Part 1, 1911.

229. ——, *Über die Verteilung Galaktischer Objekte*, A. N., **192**, 261, 1912.

230. ——, *Über die Helligkeit der Plejadennebel*, A. N., **195**, 449, 1913.

231. ——, *Comparison between the Distribution of Energy in the Spectrum of the Integrated Light of the Globular Cluster M* 3 *and of Neighboring Stars*, Ap. J., **41**, 10, 1915.

232. ——, *N. G. C.* 1647, Mt. W. Rep., **9**, 222, 1914.

233. ——, *Photographische Sterngrössen Schwacher Zentralplejaden*, A. N., **199**, 247, 1914.

234. ——, *Effective Wave-lengths of 184 Stars in the Cluster N. G. C.* 1647, Mt. W. Contr. 100, 1915.

235. ——, *Prüfung, der photographischen Grössenskala der hellen Plejadensterne*, A. N., 200, 137, 1915.

236. ——, *Photographische Sterngrössen von 233 Praesepesternen*, A. N., 203, 261, 1916.

237. ——, *The Nature of Globular Clusters*, Obs., 40, 303, 1917.

238. ——, *Ein schwacher Verdunkelungsveränderlicher in Praesepe*, A. N., 205, 33, 1917. (The star has no sensible motion and is thus unconnected with the cluster.)

239. ——, *Photographische Sterngrössen von 308 Praesepesternen*, A. N., 205, 71, 1917.

240. ——, *Photographische Messung der Lichtverteilung im mittleren Gebiet des kugelförmigen Sternhaufens Messier* 3, A. N , 207, 89, 1918.

241. ——, *Bemerkungen zur Hyadengruppe*, A N., 209, 113, 1919.

242. ——, *Photographisch-spektralphotometrische Grossen von Hyadensternen*, A N., 209, 115, 1919.

243. ——, *Bearbeitung der J. F. J. Schmidtschen Beobachtungen und Bestimmung der Periode von δ Cephei*, A. N., 210, 17, 1919.

244. ——, *On the Motion of the Magellanic Clouds*, M. N. R. A. S., 80, 782, 1920.

245. ——, *Effective Wavelengths of Stars in the Pleiades*, Mem. Danish Acad., (8), 4, No. 4, 1923.

246. ——, *Stars possibly belonging to the group of the Hyades selected from a comparison between A. G. Berl. A and A. G. Berl. B with Abbadia* 1915, B. A. N., 1, 4, 1921.

247. ——, *Remarks on some Double Stars in the Hyades*, B. A. N., 1, 87, 1922.

248. ——, *Photographic Observations of RR Lyrae from Plates taken at Potsdam*, 1910–1913, B. A. N , 1, 139, 1922.

249. ——, *On the Motion of Praesepe and of the Hyades*, B. A N., 1, 150, 1922.

250. ——, *On the Motions of the Clusters χ and h Persei*, B. A. N., 1, 151, 1922.

251. ——, *Notes on the Magnitude Scale of the Plejades*, B. A. N , 1, 152, 1922.

252. ——, *Further Remark on the Motion of the Clusters χ and h Persei*, B. A. N., 1, 218, 1923.

253. ——, *A Star in the Pleiades possibly belonging to the System of the Hyades*, B. A. N., 3, 108, 1926.

254. ——, *The Pleiades*, M. N. R. A. S., 89, 660, 1929.

255. HINKS, A. R., *On the Galactic Distribution of Gaseous Nebulae and of Star Clusters*, M. N. R. A. S., 71, 693, 1911.

256. HINS, C. H., *B. D. +13° 688, a Star of the Hyades?*, B. A. N., 2, 60, 1924.

257. HÖFFLER, F., *Versuch einer Ermittelung der Parallaxe des Systems Ursa Major*, A. N., 144, 369, 1897.

258. HOFFMEISTER, C., *Veränderliche in Sternhaufen*, G. u. L., Appendix II, 1920.

259. HOGG, F. S., *A Spectrophotometric Study of the Brighter Pleiades. II. The Continuous Background*, H. C. 309, 1927. (See Ref. 432, below.)

260. ——, *A Method for the Photometry of Surfaces, with an Application to the Globular Cluster Messier* 13, H. B. 870, 1929.

261. HOLDEN, E. S., *Index Catalogue of Books and Memoirs Relating to Nebulae and Clusters*, Smiths. Misc. Collects., 311, 1877.

262. ———, *On Some of the Consequences of the Hypothesis, Recently Proposed, that the Intrinsic Brilliancy of the Fixed Stars is the Same for Each Star*, Proc. A. A. A. S., **29**, 137, 1880 (early photometric considerations of clusters).

263. ———, *List of New Nebulae and Clusters Discovered in the Zone Observations at the Washburn Observatory*, Publ. Wash. Obs., **1**, 73, 1881.

264. ———, *Bemerkungen zur Plejadengruppe*, A. N., **108**, 439, 1884.

265. ———, *Photograph of the Cluster M 34 = G. C. 584*, P. A. S. P., 3, 62, 1891.

266. ———, *Characteristic Forms within the Cluster in Hercules*, P. A. S. P., **3**, 375, 1891.

267. HOLETSCHEK, J., *Über den Helligkeitseindruck von Nebelflecken und Sternhaufen*, VJS. d. A. G., **33**, 270, 1898.

268. ———, *Über den Helligkeitseindruck von Sternhaufen*, Wiener Berichte, **110**, 1253, 1901.

269. ———, *Über den Helligkeitseindruck einiger Nebelflecken und Sternhaufen*, Astronomischer Kalender fur 1904.

270. ———, *Beobachtungen über den Helligkeitseindruck von Nebelflecken und Sternhaufen*, An. der K. K. Univ. Sternw., Wien, **20**, 40, 1907 (a catalogue of the visual integrated magnitudes of 603 clusters and nebulae).

271. HOPMANN, J., *Die Sternhaufen*, Naturwiss., 8, 740, 1920.

272. ———, *Photometrische Untersuchungen von Nebelflecken*, A. N., **214**, 425, 1921.

273. ———, *Der kugelförmige Sternhaufen N. G. C. 5466*, A. N., **217**, 333, 1922.

274. ———, *Über die kosmische Stellung der Kugelhaufen und Spiralnebel*, A. N., **218**, 97, 1922.

275. ———, *Die offenen Sternhaufen N. G. C. 6885 bei 20 Vulpeculae und M 36 in Auriga*, Bonn Veröff. 19, 1924.

276. ———, *Vergleich der Hamburger und Bonner Vermessungen des kugelformigen Sternhaufens N. G. C. 5466*, A. N., **229**, 209, 1927.

277. HOWE, H. A., *Observations of Nebulae Made at the Chamberlin Observatory*, M. N. R. A. S, **61**, 29, 1900 (description of N. G C. 6981).

278. HRABÁK, M., *Bestimmung des Diameters und der Anzahl der Sternen im Sternhaufen N. G. C. 7243*, Rus. Astr. Journ , 3, 30, 1926.

279. ———, *Über die effektiven Wellenlängen des offenen Sternhaufens N. G. C. 7243*, Rus. Astr. Journ., **5**, 166, 1928.

280. HUBBLE, E., *Two New Globular Clusters*, Mt. W. Rep., **16**, 233, 1920.

281. ———, *N. G. C. 6822, a Remote Stellar System*, Mt. W. Contr. 304, 1925.

282. INNES, R. T. A., and J. VOÛTE, *Some Stars with sensible Proper Motion on an Astrographic Plate centered upon ω Centaurus*, U. C. 25, 1915.

283. ——— and others, (*Estimates of Color of 47 Tucanae*), U. C. 31, 1915.

284. ———, *Centennial proper motions of Stars near ω Centaurus*, U. C. 45, 1919.

285. ———, *Nebulae, Clusters, etc., on Sydney Plates*, U. C. 48, 1920.

286. ———, *Nebulae and Clusters in the Melbourne Zone*, U. C. 53, 1921.

287. ———, *Variable Stars in and near the Cluster ω Centaurus*, U. C. 59, 1923.

288. ———, *Proper Motions found and measured with the Blink Microscope, 2. Region Around ω Centaurus*, U. C 59, 1923.

289. ———, *Catalogue of Clusters and Nebulae near the Large Magellanic Cloud*, U. C. 61, 1924.

290. ———, *Globular Star Cluster, N. G. C. 5824*, U. C. 66, 1925.

291. ——— and W. H. VAN DEN BOS, *Nebulae, etc., in the Large Magellanic Cloud*, U. C. 73, 1927.

292. JACOBY, H., *The Rutherfurd Photographic Measures of the Group of the Pleiades*, N. Y. Acad. Ann., 6, 239, 1891.

293. JEANS, J. H., *On the "Kinetic Theory" of Star-clusters*, M. N. R. A. S., 74, 109, 1913.

294. ———, *On the Theory of Star-Streaming, and the Structure of the Universe*, M. N. R. A. S., 76, 552, 1916.

295. ———, *On the Law of Distribution in Star-clusters*, M. N. R. A. S., 76, 567, 1916.

296. ———, *The Dynamics of Moving Clusters*, M. N R. A. S., 82, 132, 1922.

297. ———, *The Evolution of Star Clusters*, Problems of Cosmogony and Stellar Dynamics, Chapter X, p. 220, 1919.

298. JOY, A. H., *An Investigation of the Cluster M 37 (N. G. C. 2099) for Proper Motion from Plates taken with the 40-inch Refractor of the Yerkes Observatory (Abstract)*, Pop. Astr., 23, 603, 1915.

299. ———, *An Investigation of the Cluster M 37 (N. G. C. 2099) for Proper-Motion*, A. J., 29, 101, 1916.

300. JUNG, J., *Die Radialgeschwindigkeiten von elf Plejadensternen nach Spektrogrammen von Professor Hartmann*, Göttingen Mitt., 17, 1914.

301. KAPTEYN, J. C., W. DE SITTER, and A. DONNER, *Parallaxes of the Clusters h and χ Persei*, Groningen Publ. 10, 1902.

302. ———, ———, and ———, *The Proper Motion of the Hyades*, Groningen Publ. 14, 1904.

303. ———, *Moving Clusters*, Trans. Internat. Solar Union, 3, 215, 1911.

304. ——— and P. J. VAN RHIJN, *The Proper Motions of δ Cephei Stars and the Distances of the Globular Clusters*, B. A. N., 1, 37, 1922.

305. KEELER, J. E , *Photographs of Nebulae and Clusters made with the Crossley Reflector*, Lick Publ., 8, 1908.

306. KEMPF, P., *Beobachtungen von Nebelflecken und Sternhaufen mit einem Lamellen mikrometer*, A. N., 129, 233, 1891.

307. ———, *Beobachtungen von Nebelflecken und Sternhaufen*, Pots Publ , 8, No 29, 147, 1892.

308. KIENLE, H., *Die Absorption des Lichtes im Interstellaren Raume*, Jahrb. d. Radioaktivität und Elektronik, 20, 1923.

309. ——— and P. TEN BRUGGENCATE, *Die Absolute Helligkeit der Plejadensterne*, Zeit. f. Phys., 28, 373, 1924. (Kienle replies to criticism of this note in Zeit. f. Phys., 31, 605, 1924.)

310. ———, *Die Gestalt der Kugelförmigen Sternhaufen*, Naturwiss., 15, 243, 1927.

311. ———, *Zur Entfernungsbestimmung von Sternsystemen*, A. N., 230, 243, 1927.

312. ——, *Die Dichteverteilung in Ellipsoidformigen Sternhaufen*, A. N., **232**, 427, 1928.

313. ——, *Die Absorption des Lichtes und die Grenze des Sternsystems*, Zs. f. Phys., **20**, 338, 1924.

314. KLEIN WASSINK, W. J., *Stars belonging to the Cluster Praesepe*, B. A. N., **2**, 183, 1924.

315. ——, *The Proper Motion and the Distance of the Praesepe Cluster*, Groningen Publications, 41, 1927.

316. KNOBEL, E. B., *Reference Catalogue of Astronomical Papers and Researches. 4, Nebulae and Clusters*, M. N. R. A. S., 36, 365, 1876.

317. KOBOLD, (*Ursa Major System*), Der Bau des Fixsternsystems, p. 145, 1906.

318. KOHLMANN, A. F., *Star Clusters: some Observations and Comparisons*, Monthly Reg. Soc. Prac. Astr., 8, 25, 1916.

319. KOHLSCHÜTTER, A., *Die Spektren der Hyaden und der Praesepe*, A. N., **211**, 289, 1920.

320. KÖNIG, A., *Photographische Vermessung der Plejaden*, A. N., **222**, 177, 1924.

321. KOPFF, A., *Über die Haufigkeitsfunktion beim Kugelsternhaufen M* 13, A. N., **219**, 311, 1923.

322. KOSTINSKY, S., *Über die Eigenbewegung der Sterne in der Umgebung der Sternhaufen χ und h Persei*, A. N. 4366.

322a. ——, *Durchmusterung der Eigenbewegungen in der Umgebung der Sternhaufens N. G. C. 7209*, A. N., **238**, 245, 1930.

323. KOSTITZIN, W., *Die Struktur kugelförmiger Sternhaufen*, Moscow Obs., **1**, 28, 1922 (in Russian).

324. ——, *Sur la Distribution des Etoiles dans les Amas Globulaires*, Bul. Astr., **33**, 289, 1916.

325. ——, *Sur la Structure des Systèmes Stellaires*, Rus. Astr. Journ., **3**, 1, 1926 (in Russian).

326. KREIKEN, E A, *Proper Motions of Stars belonging to the Plejades*, B. A. N., **2**, 55, 1924.

327. KRETZ, W. C, *The Positions and Proper Motions of the Principal Stars in the Cluster of Coma Berenices, as deduced from Measurements of the Rutherford Photographs*, Columbia Contr. No. 16, 1900.

328. KRITZINGER, H. H., *Beobachtungen der Helligkeit einiger Nebel und Sternhaufen*, Sirius, **48**, 111, 1915.

329. KRUSE, W., *Mikrometrische Vermessung des Sternhaufens N. G. C. 6633*, Heidelberg Veröff., 7, No. 3, 55, 1913.

330. KÜSTNER, F., *Der kugelförmige Sternhaufen Messier 56*, Bonn Veröff, 14, 1920.

331. ——, *Der kugelförmige Sternhaufen Messier 15*, Bonn Veröff., 15, 1921.

332. ——, *Der kugelformige Sternhaufen Messier 3*, Bonn Veröff., 17, 1922.

333. ——, *Ausmessungen der vier offenen Sternhaufen N. G. C. 7789, Messier 11 und 35, und N. G. C. 6939*, Bonn Veroff., 18, 1923.

333a. LAGRULA, J., *Études sur les Occultations d'Amas d'Étoiles par la Lune*, Ann. de l'Université de Lyon, N. S. 1, 5.

334. LAIS, G., *Esplorazione astrofotografica del Gruppo di Stelline M 52 nella costellazione di Cassiopeia*, Specola Vaticana, 1, Part 3, 1925

335. LAKITZ, F., *Csillaghalmazoc (Sternhaufen und Nebel)*, Term. Kös., 40, 1908 (Magyar).

336. LAMONT, J. VON, *Der Sternhaufen h Persei*, München Obs., Ann., 17, 1869.

337. LAMPLAND, E. O., *N. G. C.* 2419, H. B. 776, 1922.

338. LARINK, J, *Die Veranderlichen Sterne im Kugelsternhaufen Messier 3*, Bergedorf Abh., 2, No. 6, 1922.

339. ———, *Neuer Veranderlicher 4, 1921 Canum Venaticorum vom XX Cygni-Typus im Sternhaufen Messier 3*, A. N , **214**, 71, 1921.

340. ———, *Die kugelformigen Sternhaufen*, Weltall, **25**, 18, 1925.

341. LEAVITT, H. S, 105 *New Variable Stars in Scorpius*, H. C. 90, 1904. (Thirty-three of these are in Messier 4.)

342. ———, *New Variable Stars in the Small Magellanic Cloud*, Amer. Astr. Soc., **1**, 257, 1905.

343. ———, 1777 *Variables in the Magellanic Clouds*, H. A., 60, No. 4, 1908.

344. ———, *Periods of Variable Stars in the Small Magellanic Cloud*, Amer. Astr. Soc., **2**, 62, 1911.

345. ———, *Periods of 25 Variable Stars in the Small Magellanic Cloud*, H. C. 173, 1912.

346. LEDERSTEGER, K., *Das System der Bärensterne*, A. N., **224**, 153, 1924.

347. LEE, O. J., *Motions of Stars in the Scattered Cluster N. G. C.* 225, M. N. R. A. S., **86**, 645, 1926.

348. LENSE, J., *Sternbewegungen in ellipsoidisch geschichteten Sternhaufen*, A. N , **204**, 17, 1916.

349. LINDBLAD, B., *Note on the Distances of the Cluster-Type Variables*, Ap. J., **59**, 37, 1924.

350. ———, *On the Dynamics of the System of Globular Clusters*, Ups. Medd. 4, 1926.

351. LINDEMANN, E., *Helligkeitsmessungen der Bessel'schen Plejadensterne*, St. Petersburg Acad. Sci., Mem , **32**, No. 6, 1884.

352. ———, *Helligkeitsmessungen im Sternhaufen h Persei*, St. Petersburg Acad. Sci., Bul , **5**, No. 2, 55, 1895.

353. LOHSE, O., *Uber die photographische Aufnahme Sternhaufens χ Persei*, A. N., **111**, 147, 1885; **115**, 9, 1886.

354. LÖNNQUIST, C., *On the Evolution of the Stars with Mass Reduction*, Ark. Mat. Astr. o. Fys., 20A, No. 21, 1927.

355. LOUS, K., *De kugleformede stjernehobe*, Nord. Astr. Tid., **1**, 65, 1920.

356. ———, *Nogle aabne stjernegrupper*, Nord. Astr. Tid., **3**, 53, 1922.

357. LUDENDORFF, H., *Der grosse Sternhaufen im Herkules M 13*, Pots. Publ , **15**, No. 50, 1, 1905.

358. ———, *Nachtrag zu der Abhandlung, "Der grosse Sternhaufen im Herkules M 13,"* A. N., **178**, 369, 1908.

359. ———, *Bemerkungen zu Herrn Zurhellens Abhandlung "Der Sternhaufen Messier 46,"* A. N., **182**, 219, 1909.

360. ———, *Uber die Radialgeschwindigkeiten von β, ε, ζ Ursae Majoris und über die Bewegung und Parallaxe der sieben Hauptsterne des Grossen Bären*, A. N., **180**, 265, 1909.

361. LUNDBORG, A., *Nebulosor och stjarnhopar*, Pop. Astr. Tid., **4**, 16, 1923.

362. LUNDMARK, K., and B. LINDBLAD, *Photographisch effektive Wellenlängen fur einige Spiralnebel und Sternhaufen*, A. N., **205**, 161, 1917.

363. ———— and ————, *Photographic Effective Wave-Lengths of Nebulae and Clusters; Second Paper*, Ap. J., **50**, 376, 1919.

364. ————, *The Relations of the Globular Clusters and Spiral Nebulae to the Stellar Syst..n*, Swedish Acad. Proc., **60**, No. 8, 1920.

365. ————, *The Parallax of the Coma Berenices Cluster*, L. O. B., **10**, 149, 1922.

366. ————, *Die Stellung der kugelförmigen Sternhaufen und Spiralnebel zu unserem Sternsystem*, A. N., **209**, 369, 1919.

367. ————, *Determination of the Apex of Globular Clusters*, P. A. S. P., **35**, 318, 1923.

368. ————, *Avlägsna Stjärnsystem* (*Tre Extragalaktiska System nu bekanta*), Pop. Astr. Tid., **5**, 54, 1924 (the Magellanic Clouds, N. G. C. 4449, and Messier 82)

369. ————, *Plejadernas Avstånd*, Nord. Astr. Tid , **5**, 73, 1924.

370. ————, *De Magalhäeska Molnens Avstånd*, Pop. Astr. Tid., **5**, 146, 1924.

371. ————, *De Mörka Nebulosornas Utbredning*, Ups. Medd. 12, 1926 (abstract in English).

371a. ————, *Are the Globular Clusters and the Anagalactic Nebulae Related?*, Pop. Astr., **38**, No. 1, p. 26, 1930; P. A. S. P., **42**, 23, 1930.

372. LUYTEN, W. J., *Note on the Cluster N. G. C.* 6633, M. N. R. A. S., **81**, 213, 1921.

373. ————, *On the Distances of the Cepheids*, P. A. S. P., **34**, 166, 1922.

374. ————, *Note on some Stars belonging to the Hyades Cluster*, P. A. S. P., **36**, 86, 1924.

375 ————, *Proper Motions of Eleven Cluster Type Variables*, H. B. 847, 1927.

376. MAANEN, A. VAN, *The Proper Motions of 1418 Stars in and near the Clusters h and χ Persei*, Utrecht, 1911.

377. ————, *Investigations on Proper Motion, First Paper. The Motions of 85 Stars in the Neighborhood of Atlas and Pleione*, Mt. W. Contr. 167, 1919.

378. ————, *Remarks on the Motion of the Stars in and near the Double Cluster in Perseus*, Pop. Astr., **25**, 108, 1917.

379. ————, *Investigations on Proper Motion, Third Paper. The Proper Motion of Stars in and near the Double Cluster in Perseus*, Mt. W. Contr. 205, 1920.

380. ————, *Note on the Parallax of Cepheid Variables*, P. A. S. P., **32**, 62, 1920.

381. ————, *Investigations on Proper Motion, Eleventh Paper. The Proper Motion of M 13 and its Internal Motion*, Mt. W. Contr. 284, 1, 1924.

382. ————, *Investigations on Proper Motion, Twelfth Paper. The Proper Motions and Internal Motions of Messier 2*, 13, 56, Ap J , **66**, 89, 1927.

383. ————, *Over de Eigenbeweging van en in de drie bolvormige Sterrehoopen Messier 13, 56 en 2*, Versl. de gewone vergadoring der Aufdeeling Natuurkunde, **36**, No. 6, 1927.

384. MACKLIN, H. E., *The Clusters h and χ Persei*, M. N. R. A. S., **81**, 400, 1921.

385. ————, *Note on the Proper Motions of Stars of the Clusters h and χ Persei*, M. N. R. A. S., **83**, 79, 1922.

386. MACPHERSON, H., *The Distances of Star-Clusters*, Obs , **42**, 126, 1919.

387. MALMQUIST, K. G., *Über die Entfernung des offenen Haufens N. G. C.* 752, Lund Medd. 111, 1926.

388. ———, *On the Zero Point of the Period-Luminosity Curve*, Lund Medd. 113, 1926.

388a. MARTENS, ERIK, *A Research on the Spherical Dynamical Equilibrium-distribution of Stars of Unequal Masses*, Göteborg, 1928.

389. McLAUGHLIN, D. B., *The Present Position of the Island Universe Theory of the Spiral Nebulae*, Pop. Astr., **30**, 286, 1922.

390. MELLOR, T. K., *The Pleiades*, J. B. A. A., **12**, 29, 1901.

391. MELOTTE, P. J., *A Catalogue of Star Clusters shown on Franklin-Adams Chart Plates*, Mem. R. A. S., **60**, 175, 1915.

392. MESSOW, B., *Die Beiden Sternhaufen im Perseus N. G. C.* 869 *und* 884, Berge-dorf Abh., **2**, No. 2, 1913.

393. MEYER, H., (N. G. C. 6802), Dissertation, Breslau, 1902.

394. ———, *Ausmessung eines Sternhaufens in der Vulpecula*, A N , **167**, 321, 1904.

395. MILLOSEVICH, E., *Sul Cluster 4318 del C.G. (M 6)*, A. N , **113**, 345, 1885.

396. MOULTON, F. R., *Some Dynamical Considerations on Globular Star Clusters*, Amer. Astr. Soc., **1**, 329, 1909.

397. MÜLLER, G , and P. KEMPF, *Bestimmung der Helligkeit von 96 Plejaden-sternen*, A. N., **150**, 193, 1899.

398. NABAKOV, M., *La Grandeur Stellaire intégrale d' Amas et de Nébuleuses*, Rus. Astr Journ , **1**, 115, 1924

399. ———, *La Distribution de l' Éclat de l' Image extrafocale de l' Amas Stellaire N. G. C 6205 (Messier 13)*, Rus. Astr. Journ , **1**, 109, 1924.

400. ———, *L' Éclat integral des Amas Stellaires*, Rus. Astr. Journ., **2**, No. 1, 66, 1925 (in Russian).

401. ———, *Ergebnisse von Beobachtungen über die Sternhaufen*, A. N., **228**, 425, 1926.

402. NANGLE, J , *The Cluster near κ Crucis*, J. B. A. A., **18**, 384, 1908.

403. ———, *The Cluster about κ Crucis*, J. B. A. A., **19**, 27, 141, 1908.

404. NAUMANN, H., *Der Sternhaufen Praesepe; Beobachtungen mit dem Helio-meter*, Leipzig Veröff. No 2, 1927.

405. NIJLAND, A. A., N. G. C 6633, Dissertation, Utrecht, 1897.

406. ———, *Ausmessung des Sternhaufens G. C.* 4410, A. N., **144**, 257, 1897.

407. ———, *De boun van het Heelal*, Haarlem, 1924.

408. NORDLUND, J. O., *Photographische Ausmessung des Sternhaufens Messier 37*, Ark. Mat. Astr. o. Phys., **5**, No. 17, 1909.

409. NORLIND, W., *Beobachtungen einiger Sternhaufen mit dem Steinheilschen Äquatoreale*, Medd. från Obs. Uranienborg (Barsebäck) No. 2, 1917.

410. NUŠL, F., *Über die Milchstrasse, Sternhaufen, und Nebel*, Bohemian Rev., **5**, 578, 1902 (in Bohemian).

411. O' CONNOR, J., *The Taurus Cluster*, A. J., **28**, 175, 1914.

412. OLANDER, V. R., *Die Eigenbewegungsverhältnisse des Sternhaufens in der Coma Berenices*, Soc. Scient. Fennica, Com Phys Math., **4**, 11, 1927.

413. OLSSON, K., *Photographische Ausmessung der Plejaden und des Sternhaufens G C. 1712 (M 67)*, Astr. Iakt. o. Undersök anst. på Stockholm Obs., **6**, No. 4, 1898.

414. OORT, J. H., *On a Possible Relation between Globular Clusters and Stars of High Velocity*, P. N. A. S., **10**, 256, 1924.

415. ———, *Additional Notes concerning the Rotation of the Galactic System*, B. A. N , **4**, 91, 1927.

416. PACKER, D. E., *On a New Variable Star near the Cluster 5 M Librae*, Sid. Mess., **9**, 381, 1890; Eng. Mech., **51**, 378, 1890.

417. ———, *New Variable Stars near the Cluster 5 M Librae*, Sid. Mess., **10**, 107, 1891.

418. ———, *The Variable Stars (True and False) near 5 M Librae*, Eng. Mech., **52**, 80, 1890.

419. PALMER, H. K., *The Distribution of Stars in the Cluster M 13 Herculis*, Ap. J., **10**, 246, 1899.

420. PANNEKOEK, A., *Luminosity Function and Brightness for Clusters and Galactic Clouds*, B. A. N., **2**, 5, 1923.

421. ———, *New Reduction of von Zeipel's Magnitudes in Messier 3*, B. A. N., **2**, 12, 1923.

422. PARASKEVOPOULOS, J. S., *Integrated Magnitude of 47 Tucanae*, H. B. 824, 1925.

423. ———, *Five New Variable Stars*, H. B. 813, 1925 (in region of N. G. C. 6809).

424. PARENAGO, P., *Über die Helligkeit der Sternhaufen*, Bul. Obs. Corp. 9, 60–62, 1927.

425. PARIJSKI, N., *L' Essai sur l' estimation de la masse et du nombre d' étoiles de l' Amas globulaire M 13*, Rus. Astr. Journ., **3**, 10, 1926 (in Russian).

426. PARSONS, H. McW., *Photo-Visual Magnitudes of the Stars in the Pleiades*, Ap. J., **47**, 38, 1918.

427. PARVULESCO, C., *L' Amas double de Persée et les Mouvements Propres dans la Région de cet Amas*, Bul. Astr., (Mém. et Var.), **3**, 393, 1923.

428. ———, *Sur la Distribution des Etoiles dans les Amas Globulaires M 9, M 10, M 12, et la Théorie cinétique des gaz*, C. R., **181**, 500, 1925.

429. ———, Sur les Amas globulaires d' étoiles et leurs Rélations dans l' Espace, Dissertation, 1925.

430. ———, *Les Amas Globularies d' Étoiles*, l' Astr., **41**, 49, 1927.

431. PAYNE, C. H., *Proper Motions of the Stars in the Neighborhood of M 36 (N. G. C. 1960)*, M. N. R. A. S., **83**, 334, 1923.

432. ——— and F. S. HOGG, *A Spectrophotometric Study of the Brighter Pleiades. I. The Line Intensities*, H. C. 303, 1927. (See Ref. No. 259.)

433. ———, *Color Magnitudes of Seventeen Stars Near N. G. C. 6231*, H. B. 848, 1927.

434. PEASE, F. G., *The Star Cluster N. G. C. 6760*, P. A. S. P., **26**, 204, 1914.

435. ———, *Spectra of Stars in the Hercules Cluster M 13*, P. A. S. P., **26**, 204, 1914.

436. ———, *Spectra of Stars in the Hercules Cluster M 13*, Mt. W. Rep., **9**, 219, 1914.

437. ———, *Spectra of Stars in the Hercules Cluster M 13*, Mt. W. Rep., **10**, 268, 1915.

438. ——— and H. SHAPLEY, *On the Distribution of Stars in Twelve Globular Clusters*, Mt. W. Contr. 129, 1917.

439. ———, and ———, *Note on the Elliptical Form of Messier* 13, Amer. Astr. Soc., **3**, 274, 1916.

440. ——— and ———, *Axes of Symmetry in Globular Clusters*, Mt. W. Comm. 39, 1917.

441. ———, *A Planetary Nebula in the Globular Cluster Messier* 15, P. A. S. P., **40**, 342, 1928.

442. PERRINE, C. D., *A Division of the Stars in some of the Globular Star Clusters, according to Magnitude*, Ap. J., **20**, 354, 1904.

443. ———, *Discovery of many small Nebulae near some of the Globular Star Clusters*, P. A. S. P., **20**, 237, 1908.

444. ———, *Some Results derived from Photographs of the Brighter Globular Star Clusters*, L. O. B. 155, 1909.

445. ———, *The Nature of Globular Clusters*, Obs., **40**, 166, 1917.

446. ———, *Spectroscopic Notes on Southern Clusters, Nebulae and Red Stars*, P. A. S. P., **35**, 229, 1923.

447. ———, *Distances of the Galactic Cepheids, Magellanic Clouds, and Globular Clusters*, M. N. R. A. S, **87**, 426, 1927.

448. ———, *The Luminosity of the Cepheids*, Obs., **51**, 292, 1928.

448a. ———, *The Motions and Status of the Spiral Nebulae and Globular Clusters*, A. N., **236**, 329, 1929.

449. PETFR, B., *Bestimmung der Örter von* 27 *Sternen der Plejadengruppe am Meridiankreise der Leipziger Sternwarte*, A. N., **161**, 246, 1903.

450. ———, *Monographie der Sternhaufen G C.* 4460 *und G. C.* 1440, *sowie einer Sterngruppe bei* o *Piscium*, Abh. d. Math.-Phys. Kl. d. Kgl. Sachs. Ges. d. Wiss., **15**, No. 1, 1889.

451. ———, *Bestimmung der relativen Coordinaten der Sterne A und Z im Sternhaufen bei h Persei*, A. N., **131**, 51, 1892.

452. PETTIT, H. S., *The Proper Motions and Parallaxes of* 359 *Stars in the Cluster h Persei*, Pop Astr , **27**, 671, 1919.

453. PICKERING, E. C., *Variable Star in Cluster G. C.* 3636, A. N., **123**, 207, 1889. (This is Messier 3)

454. ———, *Spectrum of Pleione*, A. N., **123**, 95, 1889.

455. ———, Variable Stars near 47 Tucanae, A. N., **125**, 129, 1894.

456. ———, *Variable Star Clusters*, H. C. 2, 1895; A. N., **139**, 137, 1895; Ap. J., **2**, 321, 1895.

457. ———, *The Cluster Messier* 5 *Serpentis*, N. G. C. 5904, A. N., **140**, 285, 1896.

458. ———, *Distribution of Stars in Clusters*, H. A., **26**, 213, 1891.

459. ———, *Measurement of Positions (Stars in Messier 5)*, H. A , **26**, 226, 1891.

460. ———, *Spectra of Stars in Clusters*, H. A., **26**, 260, 1891.

461. ———, *Variable Star Clusters*, H. C. 18, 1897; A. N., **144**, 191, 1897; Ap. J., **6**, 258, 1897.

462. ———, *Variable Star Clusters*, H. C. 24, 1898; A. N., **146**, 113, 1898; Ap. J , **7**, 208, 1898.

463. ———, *Variable Stars in Clusters*, H. C. 33, 1898; A. N., **147**, 347, 1898; Ap. J., **8**, 257, 1898.

464. ———, *Variable Stars in Clusters. Rate of Increase of Light.* H C. 52, 1900; A. N., **153**, 115, 1900; Ap. J., **12**, 159, 1900.

465. PICKERING, W. H., *The Distance of the Pleiades; the Distance of Coma Berenices*, H. C. 206, 1918.

466. PIHL, O. A. L., The Stellar Cluster χ Persei micrometrically surveyed, Christiania,1891.

467. PINGSDORF, F., *Der Sternhaufen in der Cassiopeia, Messier 52*, Dissertation, Bonn,1909.

468. PITMAN, J. H., *Parallaxes of nine Stars in the Pleiades*, Amer. Astr. Soc , 5, 401, 1926.

469. PLUMMER, H. C , *The Positions of Seventy Stars in the Cluster M 13 Herculis*, M. N. R. A. S., 65, 79, 1904.

470. ——, *On the Problem of Distribution in Globular Star Clusters*, M. N. R. A. S., 71, 460, 1911.

471. ——, *Star Clusters*, Nature, 94, 674, 1915.

472. ——, *The Distribution of Stars in Globular Clusters*, M. N. R. A. S., 76, 107, 1915.

473. ——, *An Analysis of the Magnitude Curves of the Variable Stars in Four Clusters*, M. N. R. A. S., 79, 639, 1919.

474. PLUMMER, W. E., *The Great Cluster in Hercules*, M. N. R. A. S , 65, 801, 1905.

475. PORRO, F , *Variabli sospettate, M* 3, A. N , 127, 197, 1891.

476. PRITCHARD, C., *On the Proper Motions of Forty Stars in the Pleiades, both Absolute and Relative* M. N. R. A. S., 44, 355, 1884; Mem. R. A. S., 48, Part 2, 1884.

477. PROCTOR, M., *κ Crucis*, J. B. A. A., 25, 193, 1915.

478. PROCTOR, R. A., *On the Resolvability of Star-Groups Regarded as a Test of Distance* M. N. R. A. S., 30, 184, 1870.

479. PUMMERER, P., *Der Sternhaufen G. C.* 392, Publ. Küffner'schen Sternw., 6, Part 7, 1913.

480. RAAB, S , *A Research on Open Clusters,* Lund Medd., Ser. 2, 28, 1922.

481 RABOURDIN, L , *Sur les Photographies des Nébuleuses et d' Amas d' Étoiles obtenues à l' Observatoire de Meudon,* C. R., 128, 219, 1899.

482. ——, *Photographies de Nébuleuses et d' Amas d' Étoiles,* Bul. Soc. Astr. France, 13, 289, 1899.

483. RAIMOND, J. J., Jr., *The Mean Parallax of the Hyades,* B. A. N., 3, 221, 1926.

484. ——, *Proper Motions of 12 Bright Pleiades,* B. A N., 4, 135, 1928.

485. RASMUSON, N., *A Research on Moving Clusters,* Lund Medd., Ser. 2, 26, 1921.

486. ——, *Rörliga stjärnhopar,* Pop. Astr. Tid., 3, 43, 1922.

487. ——, *A New Research on the Scorpio-Centaurus Cluster,* Lund Medd. (2), 5, 47b, 1927.

488. REBEUR-PASCHWITZ, E., *Bemerkung betreffend den Sternhaufen G. C.* 1360 (*M* 35), A. N., 131, 263, 1892.

489. REBOUL, M., *La Distance des Amas d' Étoiles,* Bul. Soc. Astr. France, 32, 130, 1918.

490. ——, *Sur les Dimensions du Système Galactique,* Bul. Soc. Astr. France, 32, 244, 1918.

491. ——, *Distances des nuages stellaires et de la Voie Lactée,* Bul. Soc. Astr. France, 33, 409, 1919.

492. VAN RHIJN, P. J., *The Proper Motions of 2088 Stars and the Motion of the Open Cluster M 67, derived from Photographic Plates taken by Prof. Dr. E Hertzsprung at the Potsdam Observatory*, Groningen Publ. 33, 1922.

493. ———, *On the Absorption of Light in Space derived from the Diameter-Parallax Curve of Globular Clusters*, B. A. N , **4**, 123, 1928.

494. ———, *De Diameters van Bolvormige Sterrenhoopen en de Absorptie van het Licht in de Ruimte*, Hem. en Damp., **26**, No. 3, 1928.

495. RITCHEY, G. W., *Astronomical Photography with the Forty-inch Refractor and the Two-foot Reflector of the Yerkes Observatory*, Yerkes Publ , **2**, Part 6, 1904. (Photographs of Messier 13, Messier 15, and central part of the Orion nebula.)

496. ———, *Notes on Photographs of Nebulae taken with the 60-inch Reflector of the Mount Wilson Observatory*, M. N. R. A. S., **70**, 647, 1910.

497. RITCHIE, M., *Observations of Barnard's Variable near Messier* 11, P. A. S. P., **32**, 61, 1920.

498. ROBERTS, I , (*Photographs of Nebulae and Clusters*), M. N. R. A. S., **47**, 24, 90, 1886; **48**, 30, 1887; **50**, 315, 1890; **51**, 441, 1891; **52**, 543, 1892; **53**, 125, 331, 357, 443, 444, 445, 1893; **54**, 370, 504, 1894; **56**, 33, 380, 1896; **58**, 392, 1898; Knowledge, **24**, 11, 1901.

499. ———, Photographs of Stars, Star Clusters, and Nebulae, **1**, 1893; **2**, 1899.

500. ROWLAND, J., *Note on the Magnitude Curves in Mr. Macklin's Paper on the Clusters h and χ Persei*, M. N. R. A. S., **81**, 407, 1921.

501. RUNGE, C., (*Spectrum of Alcyone*), A. N , **145**, 228, 1897.

502. RUSSELL, W. C., *The Coloured Cluster About κ Crucis*, M. N. R. A. S., **33**, 66, 1872.

503. SALET, P., *l' Amas double de Persée*, l' Astr., **38**, 249, 1924.

504. SANFORD, R. F., *Spectroscopy of Nebulae and Star Clusters*, Mt. W. Rep., 14, 212, 1918.

505. ———, *Radial Velocities of Clusters*, Mt. W. Rep., 15, 250, 1919.

506. ———, *Spectrum of Bailey's Variable Star No. 95 in the Globular Cluster M 3 (Abstract)*, Pop. Astr., **27**, 99, 1919.

507. SAVITSKY, P., *Proper Motions of 1168 Stars of the Cluster N. G. C. 7654 (M 52) and the Surrounding Region (Second Catalogue)*, Tashkent Publ., I, 3, 1928. (See Ref. 735.)

508. SAWYER, H. B , and H. SHAPLEY, *Photographic Magnitudes of Ninety-five Globular Clusters*, H. B., 848, 1927.

509. SCHAEBERLE, J. M., *On the Physical Structure of the Great Cluster in Hercules*, A. J., **23**, 226, 1903; Bul. Soc. Astr. France, **18**, 222, 1904.

510. SCHAUB, W., *Die Welt der Kugelsternhaufen*, Weltall, **28**, 89, 1929.

511. SCHEINER, J., *Der Grosse Sternhaufen im Hercules*, Himmel und Erde, 6, 105, 1893.

512. ———, *Über den grossen Sternhaufen im Herkules, M 13*, Abh. d. Preuss. Ak. d. Wiss., Anhang, 1892.

513. ———, *Über die Liapunow'schen Messungen im Sternhaufen Messier 13*, A. N., **132**, 203, 1893.

514. ———, *Über den Sternhaufen um theta Orionis*, A. N., **147**, 149, 1898

515. SCHILLER, K., *Photographische Helligkeiten und Mittlere Örter von* 251 *Sternen der Plejadengruppe*, A. N., **171**, 337, 1906.
516. SCHILT, J., *On the Detection of Orbital Motion in Star Clusters*, P. A. S. P., **38**, 327, 1926.
517. ————, *The Distribution of Light in the Central Part of the Globular Cluster ω Centauri*, Pop. Astr., **36**, 296, 1928; A. J., **38**, 109, 1928.
518. SCHOUTEN, W. J. A., On the Determination of the Principal Laws of Statistical Astronomy, Dissertation, Amsterdam, 1918.
519. ————, *The Parallax of Some Stellar Clusters*, Obs , **42**, 112, 1919.
520. ————, *Über die Parallaxe einiger Sternhaufen*, A. N., **208**, 317, 1919.
521. ————, *The Parallax of the Pleiades*, Obs., **42**, 240, 1919.
522. SCHULHOF, L., *Vergleich der Messungen von Ferguson*, 1863–64, mit den *Beobachtungen von Bessel*, A. N., **83**, 193, 1874.
523. SCHULTZ, H., *Note in Regard to M* 92, A. N., **66**, 47, 1865.
524. ————, Globular Clusters—Motions, Upsala, 1873.
525. ————, *Mikrometrisk Bestämning af 104 Stjernor inom teleskopiska Stjerngruppen* 20 *Vulpeculae*, Proc. Swedish Acad., Supplement, 12, Sec. 1, No. 2, 1886.
526. ————, *Beobachtungen des telescopischen Sternhaufens Gen. Cat. Nr.* 4976, A. N., **108**, 371, 1884.
527. SCHUR, W., *Die Örter der helleren Sterne der Praesepe*, Göttingen Mitt., 1895.
528. ————, *Heliometrische Bestimmung der gegenseitigen Lage der Sterne A und Z in Perseus*, A. N., **132**, 95, 1892.
529. ————, *Über die Bestimmung der Parallaxe der Sterne der Praesepegruppe durch Photographische Aufnahmen*, A. N., **137**, 221, 1895.
530. ————, *Vermessung der beiden Sternhaufen h und χ Persei mit dem sechszölligen Heliometer der Sternwarte in Göttingen*, Göttingen Mitt., 6, 1900.
531. SCHWARZSCHILD, K., and W. VILLIGER, *Aufnahmen des Sternhaufens h Persei mit Spiegeln von sehr grossem Öffnungsverhältnis*, A. N., **174**, 133, 1907.
532. ————, *Über die Räumliche Bewegung der Praesepe*, A. N., **196**, 9, 1913.
533. SČIGOLEУ, B., *Über die Verteilung der Riesensterne in Sternhaufen nach dem Color-index*, Rus. Astr. Journ., **1**, No. 2, 69, 1924.
591. SEARES, F. H., and H. SHAPLEY, *Color Variation of the Cluster Type Variable RS Bootis*, P. A. S. P., **26**, 202, 1914.
534. ———— and ————, *Distribution of Colors among the Stars of N. G. C.* 1647 *and M* 67, Mt. W. Comm. 17, 1915.
615. ————and ————, *Color Variation of the Cluster Type Variable RS Bootis* (abstract), Pop Astr., **23**, 19, 1915.
535. ————, *Color-Indices in the Cluster N. G. C.* 1647, Mt. W. Contr. 102, 1916.
536. ———— and H. Shapley, *The Variation in Light and Color of RS Boötis*, Mt. W. Contr. 159, 1918.
537. ————, *The Relation of Color Index to Spectrum in the Pleiades*, P. A. S. P., **34**, 56, 1922.
538. SEE, T. J. J., *On the Theoretical Possibility of Determining the Distances of Star Clusters*, etc., A. N., **139**, 161, 1895.
539. ————, *Measures of Double Stars in the Great Nebula and Cluster in Carina*, M. N. R. A. S., **57**, 541, 1897.

540. ———, *Dynamical Theory of the Globular Clusters and of the Clustering Power inferred by Herschel from the observed Figures of Sidereal Systems of High Order*, Proc. Amer. Phil. Soc., **51**, 118, 1912.

541. ———, *Confirmation de la Valeur de la "théorie de la capture" dans l'Évolution cosmique par les plus récentes recherches sur les amas d'étoiles*, Bul. Soc. Astr. France, **28**, 476, 1914.

542. SELGA, M., *Cúmulos estelares en movimiento*, Rev. Soc. Astr. Esp., **6**, 12, 1916.

543. SELLACK, C. S., *Photographie Südlicher Sterngruppen*, A. N., **82**, 65, 1873.

544. SHAPLEY, HARLOW, *New Variables in the Center of Messier* 3, Mt. W. Contr. 91, 1914.

545. ———, *On the Nature and Cause of Cepheid Variation*, Mt. W. Contr. 92, 1914.

546. ———, *Miscellaneous Notes on Variable Stars*, Mt. W. Contr. 99, 1915

547. ——— and M. B. SHAPLEY, *A Study of the Light Curve of XX Cygni*, Mt W. Contr. 104, 1915.

548. ———, *On the Changes in the Spectrum, Period, and Light Curve of the Cepheid Variable RR Lyrae*, Mt. W. Contr. 112, 1916.

549. ———, *Studies based on the Colors and Magnitudes in Stellar Clusters, I. The General Problem of Clusters*, Mt. W. Contr. 115, 1915.

550. *II. Thirteen Hundred Stars in the Hercules Cluster (Messier* 13), Mt. W. Contr. 116, 1915.

551. *III. A Catalogue of* 311 *Stars in Messier* 67, Mt. W. Contr. 117, 1916.

552. *IV. The Galactic Cluster Messier* 11, Mt. W. Contr. 126, 1917.

553. *V. Color-Indices of Stars in the Galactic Clouds*, Mt. W. Contr. 133, 1917.

554. *VI. On the Determination of the Distances of Globular Clusters*, Mt. W Contr. 151, 1918.

555. *VII. The Distances, Distribution in Space, and Dimensions of* 69 *Globular Clusters*, Mt. W. Contr. 152, 1918.

556. *VIII. The Luminosities and Distances of* 139 *Cepheid Variables*, Mt. W. Contr. 153, 1918.

557. *IX. Three Notes on Cepheid Variation*, Mt. W. Contr. 154, 1918.

558. *X. A Critical Magnitude in the Sequence of Stellar Luminosities*, Mt. W. Contr. 155, 1918.

559. *XI. A Comparison of the Distances of Various Celestial Objects*, Mt. W. Contr. 156, 1918.

560. *XII. Remarks on the Arrangement of the Sidereal Universe*, Mt. W. Contr. 157, 1918.

561. *XIII. The Galactic Planes in* 41 *Globular Clusters*, Mt. W. Contr. 160, 1919 (with M. B. Shapley).

562. *XIV. Further Remarks on the Structure of the Galactic System*, Mt. W. Contr. 161, 1919 (with M. B. Shapley).

563. *XV. A Photometric Analysis of the Globular System M* 68, Mt. W. Contr 175, 1920.

564. *XVI. Photometric Catalogue of* 848 *Stars in Messier* 3, Mt. W. Contr. 176, 1920 (with H. N. Davis).

565. *XVII. Miscellaneous Results*, Mt. W. Contr. 190, 1920.

566. *XVIII. The Periods and Light Curves of 26 Cepheid Variables in Messier 72,* Mt. W. Contr. 195, 1920 (with M. Ritchie).

567. *XIX. A Photometric Survey of the Pleiades,* Mt. W. Contr. 218, 1921 (with M. L. Richmond).

568. ———, *The Variations in Spectral Type of Twenty Cepheid Variables,* Mt. W. Contr. 124, 1916.

439. ——— F. G. PEASE and, *on the Distribution of Stars in Twelve Globular Clusters,* Mt. W. Contr. 129, 1917.

5. ——— W. S. ADAMS and, *Note on the Cepheid Variable SU Cassiopeiae,* Mt. W. Contr. 145, 1918.

536. ——— F. H. SEARES, and, *The Variation in Light and Color of RS Bootis,* Mt. W. Contr. 159, 1918.

569. ——— and M. B. SHAPLEY, *The Light Curve of XX Cygni as a Contribution to the Study of Cepheid Variation,* Mt. W. Comm. 14, 1915.

534. ——— F. H. SEARES and, *Distribution of Colors among the Stars of N . G. C. 1647 and M 67,* Mt. W. Comm. 17, 1915.

570. ———, *Studies of Magnitudes in Star Clusters, I. On the Absorption of Light in Space,* Mt. W. Comm. 18, 1916.

571. *II. On the Sequence of Spectral Types in Stellar Evolution,* Mt. W. Comm 19, 1916.

572. *III. The Colors of the Brighter Stars in Four Globular Systems,* Mt. W. Comm. 34, 1916.

573. *IV. On the Color of Stars in the Galactic Clouds Surrounding Messier 11,* M. W. Comm. 37, 1917.

574. *V. Further Evidence of the Absence of Scattering of Light in Space,* Mt. W. Comm. 44, 1917.

575. *VI. The Relation of Blue Stars and Variables to Galactic Planes,* Mt W. Comm. 45, 1917.

576. *VII. A Method for the Determination of the Relative Distances of Globular Clusters,* Mt. W. Comm. 47, 1917.

577. *VIII. A Summary of Results Bearing on the Structure of the Sidereal Universe,* Mt W. Comm. 54, 1918.

578. *IX. The Distances and Distribution of Seventy Open Clusters,* Mt. W. Comm. 62, 1919.

579. *X. Spectral Type B and the Local Stellar System,* Mt. W. Comm. 64, 1919.

580. *XI. Frequency Curves of the Absolute Magnitude and Color Index for 1152 Giant Stars,* Mt. W. Comm. 69, 1920.

581. *XII. Summary of a Photometric Investigation of the Globular System Messier 3,* Mt. W. Comm. 70, 1920.

582. *XIII. Variable Stars in N. G. C. 7006,* Mt. W Comm. 74, 1921.

583. ———, *A Short Period Cepheid with Variable Spectrum,* Mt. W. Comm. 21, 1916.

4. ———, W. S. ADAMS and, *The Spectrum of δ Cephei,* Mt. W. Comm. 22, 1916.

584. ———, *Discovery of Eight Variable Stellar Spectra,* Mt. W. Comm. 27, 1916.

585. ———, F. G. PEASE and, *Axes of Symmetry in Globular Clusters,* Mt. W. Comm. 39, 1917. (This is the same as Ref. 440)

586. ——, and S. B. NICHOLSON, *On the Spectral Lines of a Pulsating Star,* Mt. W. Comm. 63, 1919.

587. ——, *Photometry of Star Clusters,* Mt. W. Rep., **14**, 205, 1918.

588. ——, *Limit of Brightness in Clusters,* Mt. W. Rep., **16**, 243, 1920.

589. ——, *Spectroscopic Parallaxes of G and K Stars in Clusters,* Mt. W. Rep., **16**, 242, 1920.

590. ——, *Absolute Magnitudes of Cluster Stars,* Mt. W. Rep., **17**, 265, 1921.

591. ——, F. H. SEARES and, *Color Variation of the Cluster-Type Variable RS Boötis,* P. A. S. P., **26**, 202, 1914.

592. ——, *Note on the Color of the Faint Stars in the Orion Nebula,* P. A. S. P., **27**, 40, 1915.

593. ——, *(New Variable Stars in the Hercules Cluster),* P. A. S. P., **27**, 134, 238, 1915.

594. ——, *The Colors of Fifteen Variables in Messier* 3, P. A. S. P., **28**, 81, 1916.

595. ——, *Six Cepheids with Variable Spectra,* P. A. S. P., **28**, 126, 1916.

596. ——, *Outline and Summary of a Study of Magnitudes in the Globular Cluster Messier* 13, P. A. S. P., **28**, 171, 1916.

597. ——, *Light Elements of Variable No. 37 in Messier* 3, P. A. S. P., **29**, 110, 1917.

598. —— and H. DAVIS, *On the Variations in the Periods of Variable Stars in Messier* 3, P. A. S. P., **29**, 140, 1917.

599. —— and ——, *Messier's Catalogue of Nebulae and Clusters,* P. A. S P., **29**, 177, 1917.

600. ——, *Descriptive Notes Relative to Nine Clusters,* P. A. S. P., **29**, 185, 1917.

601. ——, *A Faint Nova in the Andromeda Nebula,* P. A. S. P., **29**, 213, 1917.

602. ——, *Note on the Magnitudes of Novae in Spiral Nebulae,* P. A. S. P., **29**, 213, 1917.

603. ——, *The Dimensions of a Globular Cluster,* P. A. S. P., **29**, 245, 1917.

604. ——, *Globular Clusters and the Structure of the Galactic System,* P. A. S. P., **30**, 42, 1918.

605. —— and H DAVIS, *Note on the Distribution of Stars in the Globular Cluster Messier* 5, P. A. S. P., **30**, 164, 1918.

606. ——, *Note on the Distant Cluster N. G. C. 6440,* P. A. S. P., **30**, 253, 1918.

607. ——, *On Radiation and the Age of Stars,* P. A. S. P., **31**, 178, 1919.

609. ——, *Nineteen New Variable Stars,* P. A. S. P., **31**, 226, 1919.

610. ——, *On the Existence of External Galaxies,* P. A. S. P., **31**, 261, 1919.

611. —— and J. C. DUNCAN, *Novae in the Andromeda Nebula,* P. A. S. P., **31**, 280, 1919.

612. ——, *Photometric Parallaxes of Nine Cepheid Variables,* P. A. S. P., **32**, 162, 1920.

613. ——, *The Local Cluster,* A. N., **213**, 231, 1921.

614. ——, *Neuer Veränderlicher 2, 1922 Coronae Australis in N. G. C. 6541,* A. N., **215**, 391, 1922.

615. ——, F. H. SEARES and, *Color Variation of the Cluster-Type Variable RS Boötis (Abstract),* Pop. Astr., **23**, 19, 1915.

616. ——, *New Variables in the Center of Messier* 3 *(Abstract),* Pop. Astr., **23**, 20, 1915.

617. ———, *On the Nature and Cause of Cepheid Variation* (*Abstract*), Pop. Astr., **23**, 20, 1915.

618. ———, *Magnitudes and Colors in the Hercules Cluster* (*Abstract*), Pop. Astr., **23**, 640, 1915.

619. ———, *The Colors of Fifteen Variables in Messier* 3, Pop. Astr., **24**, 257, 1916.

620. ———, *Discovery of Eight Variable Stellar Spectra*, Pop. Astr., **24**, 354, 1916.

621. ———, *The Colors of the Brighter Stars in Seven Globular Clusters* (*Abstract*), Pop. Astr , **25**, 35, 1917.

622. ———, *Notes on the Spectra of Cepheid Variables* (*Abstract*), Pop. Astr , **25**, 36, 1917.

623. ———, F G. PEASE and, *Note on the Elliptical Form of Messier* 13 (*Abstract*), Pop. Astr., **25**, 374, 1917.

624. ———, *Notes on Stellar Clusters* (*Abstract*), Pop. Astr., **25**, 379, 1917.

625. ——— and J. C. DUNCAN, *The Globular Cluster Messier* 22 (*N. G. C.* 6656) (*Abstract*), Pop. Astr., **27**, 100, 1919.

626. ———, *The Galactic System*, Pop. Astr., **31**, 316, 1923.

627. ———, *On the Distribution of Stars in Globular Clusters*, Obs., **39**, 452, 1916.

628. ———, *Note on the Explanation of the Absence of Globular Clusters from the Mid-galactic Regions*, Obs., **42**, 82, 1919.

629. ———, *Note on Changes in the Period and Light-Curve of the Cluster Variable SW Andromedae*, M. N. R. A. S., **81**, 208, 1921.

630. ——— and A. J. CANNON, *The Local System and Stars of Class A*, H. C. 229, 1922.

631. ———, *Notes Bearing on the Distances of Clusters*, H. C. 237, 1922.

632. ——— and A. J. CANNON, *The Distribution of Stars of Spectral Class B*, N. C. 239, 1922.

633. ——— and ———, *The Distribution of Stars of Spectral Class M*, H. C. 245, 1923.

634. ———, *On the Dwarf Variable Stars in the Orion Nebula*, H. C. 254, 1924

635. ———, *The Magellanic Clouds, I. The Distance and Linear Dimensions of the Small Cloud*, H. C. 255, 1924.

636. ———, *The Magellanic Clouds, II. The Luminosity Curve for Giant Stars*, H. C. 260, 1924.

637. ———, *The Magellanic Clouds, III. The Distance and Linear Dimensions of the Large Cloud*, H. C. 268, 1924.

638 ——— and H. H. WILSON, *The Magellanic Clouds, IV. The Absolute Magnitudes of Nebulae, Clusters, and Peculiar Stars in the Large Cloud*, H. C. 271, 1925.

639. ——— and ———, *The Magellanic Clouds, V. The Absolute Magnitudes and Linear Diameters of* 108 *Diffuse Nebulae*, H. C. 275, 1925.

640. ——— and ———, *The Magellanic Clouds, VI. Positions and Descriptions of* 170 *Nebulae in the Small Cloud*, H. C. 276, 1925.

641. ———, I. YAMAMOTO, and H. H. WILSON, *The Magellanic Clouds, VII. The Photographic Period-Luminosity Curve*, H. C. 280, 1925.

642. ——— and M. L. WALTON, *The Magellanic Clouds, VIII. Note on the Spectral Composition of the Foreground*, H. C. 288, 1925.

643. ———, *Note on Obscuring Cosmic Clouds in High Galactic Latitudes*, H. C. 281, 1925.

644. ——— and M. WALTON, *Investigations of Cepheid Variables, I. The Period-Spectrum Relation*, H. C. 313, 1927.

645. ———, *Investigations of Cepheid Variables, II. The Period-Luminosity Relation for Galactic Cepheids*, H. C. 314, 1927.

646. ———, *Investigations of Cepheid Variables, III. Cluster Type Variables and Theories of Cepheid Variation*, H. C. 315, 1927.

647. ——— and M. WALTON, *Investigations of Cepheid Variables, IV. Beta Doradus, a new Fourth Magnitude Cepheid*, H. C. 316, 1927.

648. ———, *Dimensions of Messier 3*, H. B. 761, 1921.

649. ———, *Note on the Velocity of Light*, H. B. 763, 1922.

650. ———, *Parallax of Messier 5*, H. B. 763, 1922.

651. ———, *Giant Stars near the Pleiades*, H. B. 764, 1922.

652. ———, *Spectral Classification of Faint Pleiades*, H. B. 764, 1922.

653. ———, *The Absolute Magnitude of Cluster Type Variables*, H B. 765, 1922.

654. ———, *New Cluster Type Variable in the Small Magellanic Cloud*, H. B. 765, 1922.

655. ———, *Star of High Velocity*, H. B. 773, 1922. (RZ Cephei.)

656. ———, *Group of New Globular Clusters*, H. B. 775, 1922.

657. ———, *Approximate Distance and Dimensions of the Large Magellanic Cloud*, H. B. 775, 1922.

658 ———, *New Globular Clusters*, H B. 776, 1922.

659. ———, *N. G. C. 2419*, H. B. 776, 1922.

660. ———, *New Faint Cluster Variable (near N. G. C. 6362)*, H. B. 777, 1922, (variable discovered by S. I. Bailey.)

661. ———, *Absolute Magnitude of RZ Cephei*, H. B. 778, 1922.

663. ———, *Five New Variable Stars*, H. B. 781, 1923 (variables discovered by A. J. Cannon, I. E. Woods, and S. I. Bailey).

664. ———, *Globular Cluster Containing Long Period Variables*, H. B 783, 1923.

665. ———, *Nine New Variables in High Galactic Latitude*, H. B. 791, 1923.

666. ———, *Note on the Distance of N. G. C. 6822*, H. B 796, 1923.

667. ———, *Angular Dimensions of Magellanic Clouds*, H. B. 796, 1923.

668. ———, *Photographic Magnitudes in Messier 13*, H. B. 797, 1924.

669. ———, *Proof of Variability of Fourteen Stars in Orion Nebula*, H. B. 803, 1924.

670. ———, *Note on a Star Cloud in Sagittarius*, H. B. 804, 1924.

671. ———, *Comparison of Messier 33 and the Large Magellanic Cloud*, H. B. 816, 1925

672. ———, *The Absorption of Light in Space*, H. B. 841, 1926.

673. ——— and H. B. SAWYER, *The Galactic Cluster N. G. C. 6231*, H. B. 846, 1927.

508. ———, H. B. SAWYER, and, *Photographic Magnitudes of Ninety-five Globular Clusters*, H. B. 848, 1927.

674. ———, *The Distance of Messier 22*, H. B. 848, 1927.

675. ——— and H. B. SAWYER, *A Classification of Globular Clusters*, H. B. 849, 1927.

676. ——, *The Periods of Seventy-three Variables in Messier 5*, H. B. 851, 1927.

677. —— and H. B. SAWYER, *Apparent Diameters and Ellipticities of Globular Clusters*, H. B. 852, 1927.

678. —— and A. AMES, *The Coma-Virgo Galaxies, I. On the Transparency of Inter-galactic Space*, H. B. 864, 1929.

679. ——, *Relation of Apparent Magnitude to Angular Diameter for Globular Clusters*, H. B. 864, 1929.

680. —— and H. B. SAWYER, *The Distances of Ninety-three Globular Star Clusters*, H. B. 869, 1929.

680a. ——, *The Mass-spectrum Relation for Giant Stars in the Globular Cluster Messier 22*, H. B. 874, 1930.

681. ——, *Note on the Problem of Great Stellar Distances*, P. N. A. S., **8**, 69, 1922.

682. ——, *On the Relative Velocity of Blue and Yellow Light*, H. Repr. 5, 1923.

683. —— and A. J. CANNON, *Summary of a Study of Stellar Distribution*, H. Repr. 6, 1924.

684. ——, *The Distribution of the Stars*, H. Repr. 8, 1924.

685. ——, *The Magellanic Clouds*, H. Repr. 25, 1925.

686. ——, *Studies of the Galactic Center, I. The Program for Milky Way Variable Stars*, H. Repr. 51, 1928.

687. —— and H. H. SWOPE, *Studies of the Galactic Center, II. Preliminary Indication of a Massive Galactic Nucleus*, H. Repr. 52, 1928.

688. ——, *Studies of the Galactic Center, III. The Absolute Magnitudes of Long Period Variables*, H. Repr. 53, 1928.

689. ——, *Star Clusters and the Structure of the Universe*, Scientia, **26**, 269, 353, 1919; **27**, 93, 185, 1920.

690. ——, *The Size of the Galaxy*, Scientific American Monthly, **4**, 197, 339, 1921.

691. ——, *The Scale of The Universe*, Bul. Nat. Res. Coun., **2**, 171, 1921.

692. ——, *The Magellanic Clouds*, Festschrift fur Hugo von Seeliger, p. 438, 1924.

693. ——, *Globular Clusters, Cepheid Variables, and Radiation*, Nature, **103**, 25, 1918.

694. ——, *Star Clusters and their Contribution to the Knowledge of the Universe*, Proc. Amer. Phil. Soc., **58**, 337, 1919.

695. ——, *Star Cluster*, Encyclopaedia Brittanica, 1929.

696. SHAPLEY, M. B., *The Color-Curve of XZ Cygni*, Mt. W. Contr. 128, 1917.

697. SHAW, H. K, *Note on the Nebulae and Star Clusters shown on the Franklin-Adams Plates*, M. N. R. A. S., **76**, 105, 1915.

698. SHILOW, M., *Grössenbestimmung der Sterne im Sternhaufen 20 Vulpeculae*, Bul. St. Petersb. Acad. Sci. 2nd Series, **5**, No. 3, 243, 1895.

699. SILBERSTEIN, L., *The Radial Velocities of Globular Clusters and de Sitter's Cosmology*, Nature, **113**, 350, 1924.

700. ——, *The Curvature of de Sitter's Space-time derived from Globular Clusters*, M. N. R. A. S, **84**, 363, 1924.

701. ——, *New Determination of the Curvature Radius of Space-time*, Nature, **124**, 170, 1929.

702. SITTERLY, B. W., *On the Distance and Motion of the Cluster Praesepe*, A. J., **34,** 1, 1921.

703. SLIPHER, V. M., *On the Spectrum of the Nebula in the Pleiades*, Lowell Obs. Bul. 55, 1912.

704. ——, *Spectrographic Observations of Nebulae*, Pop. Astr., **23,** 21, 1915.

705. ——, *Radial Velocities of Star Clusters*, J. R. A. S. C , **11,** 335, 1917.

706. ——, *Spectrographic Observations of Nebulae and Star Clusters*, Pop. Astr., **25,** 36, 1916.

707. ——, *Spectrographic Observations of Nebulae and Star Clusters*, P. A. S. P., **28,** 191, 1916.

708. ——, *Spectrographic Observations of Star Clusters*, Pop. Astr., **26,** 8, 1918.

709. ——, *Further Notes on Spectrographic Observations of Nebulae and Clusters*, Pop. Astr , **30,** 9, 1922.

710. ——, *The Radial Velocity of Additional Globular Star Clusters*, Pop. Astr., **32,** 622, 1924.

711. SMART, W. M., *Proper Motions of Stars in the Pleiades*, M. N. R. A. S , **81,** 536, 1921.

712. ——, *The Proper Motion of the Cluster N. G. C.* 2168 (*M* 35), M. N. R. A. S., **85,** 257, 1925.

713. SMITH, A , *The Pleiades Cluster*, Engl. Mech., **74,** 469, 1901.

714. ——, *The Double Cluster in Perseus*, Engl. Mech., **75,** 94, 1902.

715. SMITH, M. F., *A Second Determination of the Relative Positions of the Principal Stars in the Group of the Pleiades*, Trans. Yale Obs., 1, Part 8, 1896.

716. SOLÁ, J. C., *Triangulation Micrométrique de l'Amas* 6523 (*M* 8), A. N., **148,** 97, 1898.

717. ——, *Observación estereoscópica del Cúmulo estelar ω del Centauro*, Bol. del Obs. Fabra, 9, 1924.

718. SPRAGUE, R., *Star Clusters*, Pop. Astr , **1,** 407, 1894.

719. STEAVENSON, W. H , *The Star Cluster N. G. C.* 2632, J. B. A. A., **26,** 265, 1916.

720. STEENWIJK, J. E. DE VOS VAN, *A Remarkable Cluster of Stars*, Obs , **42,** 315, 1919. (This is a wide scattering of peculiar stars, not strictly a cluster)

721. STEPANOFF, W., *On the Steady Spherical Star-Clusters*, Rus. Astr. Journ., **5,** 132, 1928.

722. STONE, E. J., *On a Cause for the Appearance of Bright Lines in the Spectra of Irresolvable Star Clusters*, M. N. R. A. S., **38,** 106, 484, 1878.

723. ——, *Note on the Effects of Distance upon the Spectra of Physical Clusters of Stars*, M. N. R. A. S , **57,** 9, 1896.

724. STRATONOFF, W., (*Anzahl der Plejadensterne auf Photographischen Aufnahmen*), A. N., **141,** 103, 1896.

725. ——, *Note Sur les Pléiades*, A. N., **144,** 137, 1898.

726. ——, *Amas Stellaire de l'Ecu de Sobieski* (*Messier* 11) *d' après des Mesures Photographiques*, Tashkent Publ., 1, 1, 1895.

727. ——, *Photographie à Pose longue de h et χ de Persée*, A. N., **155,** 215, 1900.

728. STRATTON, F. J. M., *Proper Motions of Faint Stars in the Pleiades*, Mem. R. A. S., **57,** Part 4, 1908.

729. STRÖMBERG, G., *Analysis of Radial Velocities of Globular Clusters and non-Galactic Nebulae*, Mt. W. Contr. 292, 1924.

730. STRÖMGREN, E., *Om Bevargelsesmulighederne I Stjernhobe*, Nord. Astr. Tid., **6**, 21, 1925.

731. ――――, *Über Bewegungsformen in Globular Clusters*, A. N., **203**, 17, 1916.

732. ―――― and B. DRACHMANN, *Über die Verteilung der Sterne in Kugelformigen Sternhaufen mit besonderer Rücksicht auf Messier 5*, Kopenhagen Publ., 16, 1914.

733. SUBBOTIN, M., *A Catalogue of the Photographic Magnitudes of 194 Stars in Messier 67*, Rus. Astr. Journ., **2**, No. 3, 47, 1925.

734. ――――, *On the Photographic Magnitudes of the Stars in the Open Cluster M 67*, A. N., **226**, 79, 1925.

735. ――――, *Proper Motions of 1186 Stars of the Cluster N. G. C. 7654 (M 52) and the Surrounding Region, First Catalogue*, Tashkent Publ., 1927. (See Ref. No. 507.)

736. SULLIVAN, R., *Star Clusters*, Pop. Astr., **26**, 432, 1918.

737. SWIFT, L., *Nebulae and Clusters*, Pop. Astr., **1**, 369, 1894.

738. TEBBUTT, J., *Note on the Probable Disappearance of Two Stars of the 6th Magnitude from the Cluster near B. A. C. 2694*, M. N. R. A. S., **35**, 126, 1874.

739. TROUVELOT, S., *Drawings of the Clusters in Hercules, M 13 and 92*, H. A., **8**, Part 2, 1876.

740. TRUMPLER, R., *Die Relativen Eigenbewegungen der Plejadensterne*, A. N., **200**, 217, 1914.

741. ――――, *Preliminary Results on the Constitution of the Pleiades Cluster*, Pop. Astr., **26**, 9, 1918.

742. ――――, *A Study of the Pleiades Cluster*, P. A. S. P., **32**, 43, 1920.

743. ――――, *The Physical Members of the Pleiades Group*, P. A. S. P., **33**, 214, 1921; L. O.B., **10**, 110, 1921.

744. ――――, *Comparison and Classification of Star Clusters*, Publ. Allegheny Obs., **6**, 45, 1922.

745. ――――, *The Cluster Messier 11*, L O. B., **12**, 10, 1925.

746. ――――, *Spectral Types in Open Clusters*, P. A. S. P., **37**, 307, 1925.

747. ――――, *Note to Mr. Doig's Letter on Spectral Types in Open Clusters*, P. A. S. P. **38**, 114, 165, 1926.

748. ――――, *B-Type Stars with Bright Hydrogen Lines in the Cluster χ Persei*, P. A. S. P., **38**, 350, 1926.

749. ――――, *Bright Line Stars in the Cluster χ Persei*, Obs., **50**, 93, 1927.

749a. ――――, *Diameters and Distances of Open Star Clusters*, P. A. S. P., **41**, 249, 1929.

750. ――――, *Magnitudes, Spectral Types and Radial Velocities in the Open Cluster Messier 39 (N. G. C. 7092)*, P. A. S. P., **40**, 265, 1928.

750a. ――――, *Preliminary Results on the Distances, Dimensions, and Space Distributions of Open Star Clusters*, L. O B. **14**, 154, 1930.

751. TURNER, H. H., *Some Measures of Photographs of the Pleiades at the Oxford University Observatory*, M. N. R. A. S., **54**, 489, 1894.

752. ――――, *Note on the Changes of Period in the Variable Bailey No. 33 in the Cluster M 5*, M. N. R. A. S., **80**, 640, 1920.

753. ———, *Further Note on Barnard's Observations of Variable Bailey No. 33 in the Cluster M5, with a Suggestion that the Comparison Star k is a Short-Period Variable*, M. N. R. A. S., **81**, 74, 1921.

754. TRZCINSKI, P., *Mglawice i zbiorowiska gwiardowe (Nebulae and Clusters)*, Wsz. **21**, 193, 1902 (in Polish).

755. VALENTINER, W., *Ausmessung des Sternhaufens G. C. 4410*, Astr. Beob. Mannheim, 3, 1879. (This is N. G. C. 6633.)

756. VALIER, M., *Das Rätsel der kugelförmigen Sternhaufen*, Astr. Zeit., **13**, 62, 1919.

756a. VANDERLINDEN, H. L., *Longueurs d'Onde Effectives des Étoiles de l'Amas de Praesepe*, Belg. Acad. Mém. Series II, **10**, Part 2, 1929.

757. VOGEL, H., *Spektra einiger Nebelflecken und Sternhaufen*, Bothkampf Beob , 1, 156, 1872.

758. ———, Der Sternhaufen χ Persei, Leipzig, 1878.

759. ———, *Muthmassliche starke Eigenbewegung eines Sterns im Sternhaufen G. C. 4440*, A N , 116, 257, 1887.

760. VOGT, H , *Photometrische Untersuchungen und Helligkeitsbestimmungen in den Sternhaufen h und χ Persei*, Heidelberg Veröff , 8, No. 3, 1921.

761. ———, *Photometrische Vermessung des Sternhaufens G. C. 1119 (M 38)*, A. N., **212**, 73, 1920.

762. ———, *Photometrische Vermessung des Sternhaufens N. G. C. 6633*, A. N., **216**, 373, 1922

763. ———, *Flächenhelligkeiten von Nebelflecken und Sternhaufen*, A. N., **221**, 11, 1924.

764. ———, *Photometrische Vermessung der Sternhaufen N. G. C. 752 und I. C. 4665*, A. N , **221**, 41, 1924.

765. VORONTSOV-VELYAMINOV, B., *Catalogue of Integrated Magnitudes of Star Clusters*, A. N., **226**, 195, 1925.

766. ———, *Integral Magnitudes of South Star Clusters*, A. N., **228**, 325, 1926.

767. ———, *Photographic Magnitudes of Globular Clusters*, A. N., **236**, 1, 1929.

767a. VYSSOTSKY, A., *Some Results of a Photometric Study of the Double Cluster in Perseus*, Pop. Astr., 36, 350, 1928.

768. WALLENQUIST, Å., *Stjarnhopen Messier 36*, Nord. Astr. Tid., 8, 140, 1927.

769. ———, *Colors and Magnitudes in the Open Cluster Messier 36*, N. G. C. 1960, Upsala Medd., 32, 1927.

770. ———, *A Research based on the Bolometric Magnitudes in the Cluster Messier 37 (N. G. C. 2099)*, Upsala Medd., **36**, 1928.

771. ———, *Om Fotometriska Undersokningar av öppna och klotformiga stjarn-hopar*, Pop. Astr. Tid., 9, 17, 1928.

772. ———, *A Photometric Research on Two Open Clusters in Cassiopeia (Messier 52 and N. G. C. 663)*, Upsala Medd., **42**, 1929.

773. ———, *A Photometric Investigation of the Open Cluster Messier 35 (N. G. C. 2168)*, Bosscha Ann., 3B, 1929.

773a. ———, *On the masses of the stars in stellar clusters and their relation to the theory of Eddington*, Proc. Fourth Pac. Sci. Cong , Java, 1929.

774. WATERS, S., *The Distribution of the Clusters and Nebulae*, M. N. R. A. S., 33, 558, 1873.

775. ——, *On Two Distribution Maps of the Nebulae and Clusters in Dr. Dreyer's Catalogue of 1888*, M. N. R. A. S., **54**, 526, 1894.

775a. WEGNER, U., *Über die Verteilungsfunktion in Kugelsternhaufen*, Zeit. f. Phys., **49**, 386, 1928.

776. WILSON, F., *Clusters and Nebulae Visible with Small Optical Means*, J. B. A. A., **27**, 72, 1916.

777. WILSON, H. C., *The Number and Distribution of the Stars in the Vicinity of the Pleiades*, Pop. Astr., **15**, 193, 1907.

778. ——, *The Hyades Group of Stars*, Pop. Astr., **20**, 359, 1912.

779. ——, *Radial Velocity of the Praesepe Cluster from Objective-Prism Neodymium Plates*, Pop. Astr., **31**, 93, 1923.

780. WILSON, R. E., *On the Radial Velocities of Five Nebulae in the Magellanic Clouds*, P. N. A. S., **1**, 183, 1915.

781. ——, *The Radial Velocity of the Greater Magellanic Cloud*, Lick Publ., **13**, 185, 1918.

782. ——, *The Proper Motions and Mean Parallax of the Cepheid Variables*, A. J., **35**, 35, 1923.

783. WINLOCK, A., *Positions of Stars in Globular Clusters*, H. A., **38**, 235, 1901.

784. WINNECKE, F. A. T., *On the Visibility of Stars in the Pleiades to the Naked Eye*, M. N. R. A. S., **39**, 146, 1878

785. WIRTZ, C., *Sternhaufen, Nebelflecke und Weltraum*, Astr. Schr. d. Bundes d. Sternfreunde, 1, 1922.

786. ——, *Einiges zur Statistik der Radialbewegungen von Spiralnebeln und Kugelsternhaufen*, A. N , **215**, 349, 1922.

787. ——, *Triangulation der Hyaden-Gruppe*, A. N., **160**, 17, 1902.

788. ——, *Flächenhelligkeiten von 566 Nebelflecken und Sternhaufen nach photometrischen Beobachtungen am 49-cm Refraktor der Universitätssternwarte Strassburg, 1911–1916*, Lund Medd., Ser. 2, 29, 1923.

789. ——, *Einiges zur Statistik der kugelförmigen und offenen Sternhaufen*, A. N., **220**, 293, 1924.

790. ——, *Totalhelligkeit des Kugelsternhaufens M* 3 = *N. G. C.* 5272, Kiel Publ., **15**, 59, 1927.

791. WOLF, M., *Die Aussen-Nebel der Plejaden*, Münch. Abh., **20**, Part 3, 615, 1900.

792. ——, *The Inner Nebulae of the Pleiades*, Knowledge and Scientific News, **1**, 288, 1904.

793. ——, *Photographische Messung der Sternhelligkeiten im Sternhaufen G C.* 4410, A. N., **126**, 297, 1890.

794. ——, *Bewegte Sterne in der Umgebung der Plejaden*, A. N., **218**, 81, 1922.

795. ——, *Die Sternleeren bei Messier* 11 *Scuti*, A. N., **229**, 1, 1927.

796. WOOD, H. E., *Note on Southern Star Clusters*, U. C. **75**, 444, 1927.

797. WOODS, I. E., *Variable Stars in the Cluster, N. G. C.* 3201, H. C. 216, 1919.

798. ——, *Variable Stars in the Cluster, N. G. C.* 6362, H. C. 217, 1919.

799. ——, *New Variable in N. G. C.* 6541, H. B. 764, 1922.

800. WORSSELL, W. M., (*Star Clusters and Nebulae from the Wolf-Palisa Chart No.* 76), U. C. 20, 1914.

801. YOUNG, A. S., *Rutherfurd Photographs of the Stellar Clusters h and χ Persei*, Columbia Contr. 24, 1905.

802 ZEIPEL, H. VON, *La Théorie des Gaz et les amas globulaires*, C. R., **144**, 361, 1907.

803. ———, *Catalogue de 1571 Étoiles contenues dans l'amas globulaire Messier 3*, (*N. G. C.* 5272), Ann. Paris Obs. Mem , **25**, FI, 1908.

804. ———, *Recherches sur la constitution des amas globulaires*, Proc. Swedish Acad., **51**, No. 5, 1913.

805. ———, *La Loi des luminosités dans l' amas globulaire M3*, Ark. Mat. Astr. o. Phys., **11**, No. 22, 1916.

806. ———, *Étoiles et molécules*, Scientia, **21**, 13, 1917.

807. ———, *Die Bestimmung der Massen der Sterne aus ihrer Verteilung in den Sternhaufen*, A. N., Jubiläunsnummer, **33**, 1921; Nord. Astr Tid., **2**, 102, 1921.

808. ———, *Om Stjärngruppen M 37*, Pop. Astr. Tid , **2**, 132, 1921.

809. ——— and J. LINDGREN, *Photometrische Untersuchungen der Sterngruppe M 37* (*N. G. C.* 2099), Proc. Swedish Acad., **61**, No. 15, 1921.

810. ———, *Om stjärngruppernas Natur*, Pop. Astr. Tid., **4**, 1, 1923.

811. ZINNER, E , *Untersuchungen uber die Farben und Grossen in den Sternhaufen*, Sirius, **51**, 87, 1918.

812. ZURHELLEN, W , *Der Sternhaufen Messier 46*, Bonn Veröff., **11**, 1909.

APPENDIX D

SPECIAL BIBLIOGRAPHIES FOR STAR CLUSTERS

(Numbers refer to titles in Appendix C)

Pleiades: 1, 6, 10, 52,* 68, 71, 81,* 99, 103, 116, 120, 121,* 153, 165–168, 182, 187,* 190, 197, 209–214, 226,* 230,* 233, 235, 245, 251, 253, 254, 259, 264, 292, 300, 309, 320, 326, 351, 369, 377, 390, 397, 413, 426, 432, 449, 454, 465, 468, 476, 484, 501, 515, 521, 537, 567, 651, 652, 703,* 711, 713, 715, 724, 725, 728, 740–743, 751, 777, 784, 791,* 792,* 794.

Hyades: 89, 123, 126, 241, 242, 246, 247, 249, 253, 256, 302, 319, 374, 411, 483, 778, 787.

Praesepe: 86, 110, 215, 216, 236, 238, 239, 249, 314, 315, 319, 404, 527, 529, 532, 702, 756a, 779.

Coma Berenices: 104, 109, 173, 218, 327, 365, 412, 465.

h and χ Persei: 2, 41, 42, 44, 92, 93, 105, 133, 172, 222, 225, 250, 252, 301, 322, 336, 352, 376, 378, 379, 384, 385, 392, 427, 451, 452, 466, 500, 503, 528, 530, 531, 714, 727, 748, 749, 758, 760, 767a, 801.

Ursa Major: 90, 156, 203, 227, 257, 317, 346, 360.

Catalogues of clusters: 30, 82, 127, 140, 141, 148–151, 194, 263, 270, 289, 305, 391, 480, 599.

Number and distribution: 23, 28–30, 40, 48, 70, 85, 99, 106–108, 119, 136, 143, 163, 178, 193, 229, 255, 364, 366, 429, 549, 555, 560, 562, 577, 578, 628, 680, 689, 774, 775.

Spectra: 3, 74a, 78, 142, 177, 179, 180, 259, 319, 432, 433, 435–437, 441, 446, 454, 460, 501, 504, 506, 537, 589, 652, 694, 703–710, 722, 723, 746–750, 757.

Motions: 1, 2, 15, 43, 59, 60, 63, 64, 79, 87–90, 154–157, 170, 173, 176, 195, 201, 210, 211, 227, 244, 249, 250, 252–254, 282, 284, 288, 298–300, 302, 303, 314, 315, 322, 326, 327, 347, 360, 367, 376–379, 381–383, 385, 412, 414, 415, 427, 431, 448a, 452, 476, 484–487, 492, 505, 507, 524, 532, 542, 699–702, 704–712, 728, 729, 735, 740, 750, 759, 779, 794.

Variable stars: 8, 9, 11–14, 17–22, 25–28, 30–39, 50, 55, 57, 58, 62, 65–67, 69, 72, 76, 105, 113, 114, 122, 130, 131, 133, 185, 186, 202, 238, 258, 287, 338, 339, 341, 416–418, 423, 453, 455–457, 461–464, 473, 475, 497, 506, 544, 566, 575, 582, 593, 594, 597, 598, 609, 614, 616, 619, 631, 634, 660, 664, 669, 674, 676, 738, 752, 753, 797–799.
See Table IV, I for bibliography of discoveries.

Period-luminosity curve: 99, 146, 343–345, 388, 554, 557, 565, 631, 641, 645, 646, 653, 681, 782.

*Refers to nebulosity in the Pleiades.

INDEX

INDEX TO CLUSTERS

Milton Keynes UK
Ingram Content Group UK Ltd.
UKHW010747110923
428455UK00006B/460